T0291613

DIFFERENTIAL GEOMETRY
OF THREE DIMENSIONS

DIFFERENTIAL GEOMETRY
OF THREE DIMENSIONS

By

C. E. WEATHERBURN, M.A., D.Sc., LL.D.
EMERITUS PROFESSOR OF MATHEMATICS
UNIVERSITY OF WESTERN AUSTRALIA

VOLUME I

CAMBRIDGE
AT THE UNIVERSITY PRESS
1961

CAMBRIDGE
UNIVERSITY PRESS

University Printing House, Cambridge CB2 8BS, United Kingdom

Cambridge University Press is part of the University of Cambridge.

It furthers the University's mission by disseminating knowledge in the pursuit of education, learning and research at the highest international levels of excellence.

www.cambridge.org
Information on this title: www.cambridge.org/9781316603840

© Cambridge University Press 1961

First edition 1927
Reprinted 1931, 1939, 1947, 1955, 1961
First paperback edition 2016

A catalogue record for this publication is available from the British Library

ISBN 978-1-316-60384-0 Paperback

PREFACE TO THE FOURTH IMPRESSION

THE present impression is substantially a reprint of the original work. Since the book was first published a few errors have been corrected, and one or two paragraphs rewritten. Among the friends and correspondents who kindly drew my attention to desirable changes were Mr A. S. Ramsey of Magdalene College, Cambridge, who suggested the revision of §5, and the late R. J. A. Barnard of Melbourne University, whose influence was partly responsible for my initial interest in the subject.

The demand for the book, since its first appearance twenty years ago, has justified the writer's belief in the need for such a vectorial treatment. By the use of vector methods the presentation of the subject is both simplified and condensed, and students are encouraged to reason geometrically rather than analytically. At a later stage some of these students will proceed to the study of multidimensional differential geometry and the tensor calculus. It is highly desirable that the study of the geometry of Euclidean 3-space should thus come first; and this can be undertaken with most students at an earlier stage by vector methods than by the Ricci calculus. A student's appreciation of the more general case will undoubtedly be enhanced by an earlier acquaintance with differential geometry of three dimensions.

The more elementary parts of the subject are discussed in Chapters I–XI. The remainder of the book is devoted to differential invariants for a surface and their applications. It will be apparent to the reader that these constitute a powerful weapon for analysing the geometrical properties of surfaces, and of systems of curves on a surface. The unit vector, $\mathbf{n}$, normal to a surface at the current point, plays a prominent part in this discussion. The first curvature of the surface is the negative of the divergence of $\mathbf{n}$; while the second curvature is expressible simply in terms of the divergence and the Laplacian of $\mathbf{n}$ with respect to the surface.

Extensive applications of these invariants to the geometry of surfaces are given in the second volume of this book. Applications to physical problems connected with curved surfaces have been given elsewhere* by the author.

* 1. On differential invariants in geometry of surfaces, with some applications to mathematical physics. *Quarterly Journal of Mathematics*, Vol. 50, pp. 230–69 (Cambridge, 1925).

2. On small deformation of surfaces and of thin elastic shells. *Ibid.*, Vol. 50, pp. 272–96 (1925).

3. On the motion of an extensible membrane in a given curved surface. *Phil. Mag.*, Vol. 23, pp. 573–80 (1937).

4. On transverse vibrations of curved membranes. *Phil. Mag.*, Vol. 28, pp. 632–34 (1939).

C. E. W.

UNIVERSITY OF W.A.,
    PERTH,
        WESTERN AUSTRALIA,
            22 *January*, 1947.

CONTENTS

CHAPTER III

CURVILINEAR COORDINATES ON A SURFACE. FUNDAMENTAL MAGNITUDES

CHAPTER IV

CURVES ON A SURFACE

LINES OF CURVATURE

CONJUGATE SYSTEMS

ASYMPTOTIC LINES

ISOMETRIC LINES

NULL LINES

CHAPTER V

THE EQUATIONS OF GAUSS AND OF CODAZZI

CHAPTER VI
GEODESICS AND GEODESIC PARALLELS
GEODESICS

CHAPTER VII
QUADRIC SURFACES. RULED SURFACES
QUADRIC SURFACES

CHAPTER VIII

EVOLUTE OR SURFACE OF CENTRES. PARALLEL SURFACES

SURFACE OF CENTRES

CHAPTER IX

CONFORMAL AND SPHERICAL REPRESENTATIONS. MINIMAL SURFACES

CONFORMAL REPRESENTATION

CHAPTER X

CONGRUENCES OF LINES

RECTILINEAR CONGRUENCES

CONCLUSION

FURTHER RECENT ADVANCES

INTRODUCTION

VECTOR NOTATION AND FORMULAE

SINCE elementary vector methods are freely employed throughout this book, some space may be given at the outset to an explanation of the notation used and the formulae required [*]. Vectors are denoted by **Clarendon** symbols [†]. The *position vector* **r**, of a point P relative to the origin O, is the vector whose magnitude is the length OP, and whose direction is from O to P. If x, y, z are the coordinates of P relative to rectangular axes through O, it is frequently convenient to write

$$\mathbf{r} = (x, y, z),$$

x, y, z being the resolved parts of **r** in the directions of the coordinate axes. The point, whose position vector is **r**, is referred to as " the point **r**." If **n** is a *unit* vector, that is to say a vector of unit length, and if

$$\mathbf{n} = (l, m, n),$$

then l, m, n are the *direction cosines* of **n**. The *module* or *modulus* of a vector is the positive number which is the measure of its length.

The law of **vector addition** is a matter of common knowledge. If three points O, P, Q are such that the vectors OP and PQ are equal respectively to **a** and **b**, the vector OQ is called the *sum* of **a** and **b**, and is denoted by $\mathbf{a} + \mathbf{b}$. The negative of the vector **b** is a vector with the same modulus but the opposite direction. It is denoted by $-\mathbf{b}$. The *difference* of two vectors **a** and **b** is the sum of **a** and $-\mathbf{b}$. We write it

$$\mathbf{a} - \mathbf{b} = \mathbf{a} + (-\mathbf{b}).$$

---

[*] For proofs of the various formulae the reader is referred to the author's *Elementary Vector Analysis* (G. Bell & Sons), of which Arts. 1—8, 12, 15—17, 23—29, 42—46, 49—51, 55—57 would constitute a helpful companion course of reading. (References are to the old edition.)

[†] In MS. work Greek letters and script capitals will be found convenient.

The commutative and associative laws hold for the addition of any number of vectors. Also the general laws of association and distribution for scalar multipliers hold as in ordinary algebra. Thus if p and q are scalar multipliers,

$$p\,(q\mathbf{r}) = pq\mathbf{r} = q\,(p\mathbf{r}),$$

$$(p + q)\,\mathbf{r} = p\mathbf{r} + q\mathbf{r},$$

$$p\,(\mathbf{r} + \mathbf{s}) = p\mathbf{r} + p\mathbf{s}.$$

If $\mathbf{r}$ is the position vector of any point on the straight line through the point $\mathbf{a}$ parallel to the vector $\mathbf{b}$, then

$$\mathbf{r} = \mathbf{a} + t\mathbf{b},$$

where t is a number, positive or negative. This equation is called the vector equation of the straight line.

Products of Vectors

If $\mathbf{a}$, $\mathbf{b}$ are two vectors whose moduli are a, b and whose directions are inclined at an angle θ, the **scalar product** of the vectors is the number $ab \cos \theta$. It is written $\mathbf{a} \cdot \mathbf{b}$. Thus

$$\mathbf{a} \cdot \mathbf{b} = ab \cos \theta = \mathbf{b} \cdot \mathbf{a}.$$

Hence *the necessary and sufficient condition that two vectors be perpendicular is that their scalar product vanish.*

If the two factors of a scalar product are equal, the product is called the *square* of either factor. Thus $\mathbf{a} \cdot \mathbf{a}$ is the square of $\mathbf{a}$, and is written $\mathbf{a}^2$. Hence

$$\mathbf{a}^2 = \mathbf{a} \cdot \mathbf{a} = a^2,$$

so that the square of a vector is equal to the square of its modulus.

If $\mathbf{a}$ and $\mathbf{b}$ are unit vectors, then $\mathbf{a} \cdot \mathbf{b} = \cos \theta$. Also the resolved part of any vector $\mathbf{r}$, in the direction of the unit vector $\mathbf{a}$, is equal to $\mathbf{r} \cdot \mathbf{a}$.

The *distributive law* holds* for scalar products. Thus

$$\mathbf{a} \cdot (\mathbf{b} + \mathbf{c} + \ldots) = \mathbf{a} \cdot \mathbf{b} + \mathbf{a} \cdot \mathbf{c} + \ldots,$$

and so on. Hence, in particular,

$$(\mathbf{a} + \mathbf{b})^2 = \mathbf{a}^2 + 2\mathbf{a} \cdot \mathbf{b} + \mathbf{b}^2,$$

$$(\mathbf{a} + \mathbf{b}) \cdot (\mathbf{a} - \mathbf{b}) = \mathbf{a}^2 - \mathbf{b}^2.$$

* *Elem. Vect. Anal.*, Art. 26.

Also, if we write $\mathbf{a} = (a_1, a_2, a_3),$

$$\mathbf{b} = (b_1, b_2, b_3),$$

the coordinate axes being rectangular, we have

$$\mathbf{a} \cdot \mathbf{b} = a_1 b_1 + a_2 b_2 + a_3 b_3$$

and $\mathbf{a}^2 = a_1^2 + a_2^2 + a_3^2.$

The last two formulae are of constant application.

The unit vector $\mathbf{n}$ perpendicular to a given plane is called its *unit normal.* If $\mathbf{r}$ is any point on the plane, $\mathbf{r} \cdot \mathbf{n}$ is the projection of $\mathbf{r}$ on the normal, and is therefore equal to the perpendicular p from the origin to the plane. The equation

$$\mathbf{r} \cdot \mathbf{n} = p$$

is therefore one form of the *equation of the plane.* If $\mathbf{a}$ is any other point on the plane, then $\mathbf{a} \cdot \mathbf{n} = p$, and therefore

$$(\mathbf{r} - \mathbf{a}) \cdot \mathbf{n} = 0.$$

This is another form of the equation of the plane, putting in evidence the fact that the line joining two points $\mathbf{r}$ and $\mathbf{a}$ in the plane is perpendicular to the normal.

The positive sense for a *rotation* about a vector is that which bears to the direction of the vector the same relation that the sense of the rotation of a *right-handed screw* bears to the direction of its translation. This convention of the right-handed screw plays an important part in the following pages.

Let OA, OB be two intersecting straight lines whose directions

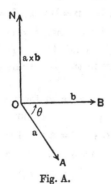

Fig. A.

are those of the two vectors $\mathbf{a}$, $\mathbf{b}$, and let ON be normal to the plane OAB. By choosing one direction along this normal as positive we fix the sense of the rotation about ON which must be regarded as positive. Let θ be the angle of rotation from OA to OB in this positive sense. Then if a, b are the moduli of $\mathbf{a}$, $\mathbf{b}$ the **vector product** of $\mathbf{a}$ and $\mathbf{b}$ is the vector $ab \sin \theta \mathbf{n}$, where $\mathbf{n}$ is the unit vector in the positive direction along the normal. This is denoted by $\mathbf{a} \times \mathbf{b}$, and is often called the *cross product* of $\mathbf{a}$ and $\mathbf{b}$. Thus

$$\mathbf{a} \times \mathbf{b} = ab \sin \theta \mathbf{n}.$$

It should be noticed that the result is independent of the choice of positive direction along the normal. For, if the opposite direction is taken as positive, the direction of $\mathbf{n}$ is reversed, and at the same time θ is replaced by $-\theta$ or $2\pi - \theta$, so that $ab \sin \theta \mathbf{n}$ remains unaltered. Hence the vector product $\mathbf{a} \times \mathbf{b}$ is a definite vector.

It is important, however, to notice that $\mathbf{b} \times \mathbf{a}$ is the negative of $\mathbf{a} \times \mathbf{b}$. For, with the above notation, the angle of rotation from OB to OA in the positive sense is $2\pi - \theta$, so that

$$\mathbf{b} \times \mathbf{a} = ab \sin (2\pi - \theta) \mathbf{n} = -\mathbf{a} \times \mathbf{b}.$$

Thus the order of the factors in a cross product cannot be changed without altering the sign of the product.

If $\mathbf{a}$ and $\mathbf{b}$ are parallel, $\sin \theta = 0$, and the cross product vanishes. Hence *the necessary and sufficient condition for parallelism of two vectors is that their cross product vanish.*

A *right-handed system* of mutually perpendicular *unit* vectors $\mathbf{t}$, $\mathbf{n}$, $\mathbf{b}$ (Fig. 3, Art. 3) is such that

$$\mathbf{t} = \mathbf{n} \times \mathbf{b}, \quad \mathbf{n} = \mathbf{b} \times \mathbf{t}, \quad \mathbf{b} = \mathbf{t} \times \mathbf{n},$$

the cyclic order of the factors being preserved throughout. We shall always choose a right-handed system of rectangular coordinate axes, so that unit vectors in the directions OX, OY, OZ satisfy the above relations.

The *distributive law* holds* also for vector products; but the order of the factors in any term must not be altered. Thus

$$\mathbf{a} \times (\mathbf{b} + \mathbf{c} + \ldots) = \mathbf{a} \times \mathbf{b} + \mathbf{a} \times \mathbf{c} + \ldots$$

and $\quad (\mathbf{b} + \mathbf{c} + \ldots) \times \mathbf{a} = \mathbf{b} \times \mathbf{a} + \mathbf{c} \times \mathbf{a} + \ldots.$

And if we write $\quad \mathbf{a} = (a_1, a_2, a_3),$

$$\mathbf{b} = (b_1, b_2, b_3),$$

* *Elem. Vect. Anal.*, Art. 28.

then, in virtue of the distributive law, and the fact that the co-ordinate axes form a right-handed system, we have

$$\mathbf{a} \times \mathbf{b} = (a_2 b_3 - a_3 b_2, \ a_3 b_1 - a_1 b_3, \ a_1 b_2 - a_2 b_1).$$

This formula should be carefully remembered.

If a vector $\mathbf{d}$ is localised in a line through the point whose position vector is $\mathbf{r}$ relative to O, the *moment* of $\mathbf{d}$ about O is the vector $\mathbf{r} \times \mathbf{d}$. Thus the moment of a vector *about a point* is a vector, sometimes called its "vector moment." It will, however, be seen shortly that the moment of $\mathbf{d}$ about an axis is a scalar quantity.

The **scalar triple product** $\mathbf{a} \cdot \mathbf{b} \times \mathbf{c}$ is the scalar product of $\mathbf{a}$ and $\mathbf{b} \times \mathbf{c}$. Except as to sign it is numerically equal to the volume of the parallelepiped whose edges are determined by the three vectors*. Its value is unaltered by interchanging the dot and the cross, or by altering the order of the factors, provided the same cyclic order is maintained. Thus

$$\mathbf{a} \cdot \mathbf{b} \times \mathbf{c} = \mathbf{a} \times \mathbf{b} \cdot \mathbf{c} = \mathbf{c} \times \mathbf{a} \cdot \mathbf{b},$$

and so on. The product is generally denoted by

$$[\mathbf{a}, \mathbf{b}, \mathbf{c}],$$

a notation which indicates the three vectors involved as well as their cyclic order If the cyclic order of the factors is altered, the sign of the product is changed. Thus

$$[\mathbf{a}, \mathbf{c}, \mathbf{b}] = -[\mathbf{a}, \mathbf{b}, \mathbf{c}].$$

In terms of the resolved parts of the three vectors, the scalar triple product is given by the determinant

$$[\mathbf{a}, \mathbf{b} \ \mathbf{c}] = \begin{vmatrix} a_1 & a_2 & a_3 \\ b_1 & b_2 & b_3 \\ c_1 & c_2 & c_3 \end{vmatrix}.$$

It is also clear that, if the three vectors $\mathbf{a}$, $\mathbf{b}$, $\mathbf{c}$ are coplanar, $[\mathbf{a}, \mathbf{b}, \mathbf{c}] = 0$; and conversely. Thus *the necessary and sufficient condition that three vectors be coplanar is that their scalar triple product vanish.*

If one of the factors consists of a sum of vectors, the product may be expanded according to the distributive law. Thus

$$[\mathbf{a}, \mathbf{b}, \mathbf{c} + \mathbf{d}] = [\mathbf{a}, \mathbf{b}, \mathbf{c}] + [\mathbf{a}, \mathbf{b}, \mathbf{d}],$$

and similarly if two or all of the factors consist of vector sums.

---

* *Elem. Vect. Anal.*, Art. 43.

The **vector triple product** $a \times (b \times c)$ is the vector product of a and $b \times c$. It is a vector parallel to the plane of b and c, and its value is given by*

$$a \times (b \times c) = a \cdot c\, b - a \cdot b\, c.$$

Similarly
$$(b \times c) \times a = b \cdot a\, c - c \cdot a\, b.$$

Both of these expansions are written down by the same rule. Each scalar product in the expansion contains the factor outside the brackets, and the first is the scalar product of the extremes.

The *scalar* product of **four vectors**, $(a \times b) \cdot (c \times d)$, is the scalar product of $a \times b$ and $c \times d$. It may be expanded† as

$$(a \times b) \cdot (c \times d) = a \cdot c\, b \cdot d - a \cdot d\, b \cdot c.$$

The *vector* product of four vectors, $(a \times b) \times (c \times d)$, may be expanded in terms either of a and b or of c and d. Thus‡

$$(a \times b) \times (c \times d) = [a, c, d]\, b - [b, c, d]\, a$$
$$= [a, b, d]\, c - [a, b, c]\, d.$$

On equating these two expressions for the product we see that *any vector d is expressible in terms of any three non-coplanar vectors a, b, c* by the formula

$$[a, b, c]\, d = [d, b, c]\, a + [d, c, a]\, b + [d, a, b]\, c.$$

If a vector d is localised in a line through the point r, its *moment about an axis* through the origin O, parallel to the *unit* vector a, is the resolved part in this direction of its vector moment about O. It is therefore equal to

$$M = a \cdot r \times d = [a, r, d].$$

Thus the moment of a vector about an axis is a scalar quantity.

The *mutual moment* of the two straight lines
$$r = a + t b,$$
$$r = a' + t b',$$

with the positive senses of the *unit* vectors b and b' respectively, is the moment about either line of the unit vector localised in the other. Thus, being the moment about the second line of the unit vector b localised in the first, it is given by

$$M = b' \cdot (a - a') \times b$$
$$= [a - a', b, b'].$$

The *condition of intersection* of two straight lines is therefore
$$[a - a', b, b'] = 0.$$

* *Elem. Vect. Anal.*, Art. 44.   † *Ibid.*, Art. 45.   ‡ *Ibid.*, Art. 46.

This is also obvious from the fact that the two given lines are then coplanar with the line joining the points **a**, **a′**, so that the vectors **b**, **b′**, **a** − **a′** are coplanar.

DIFFERENTIATION OF VECTORS

Let the vector **r** be a function of the scalar variable s, and let $\delta\mathbf{r}$ be the increment in the vector corresponding to the increment δs in the scalar. In general the direction of $\delta\mathbf{r}$ is different from that of **r**. The limiting value of the vector $\delta\mathbf{r}/\delta s$, as δs tends to zero, is called the *derivative* of **r** with respect to s and is written

$$\frac{d\mathbf{r}}{ds} = \operatorname*{Lt}_{\delta s \to 0} \frac{\delta\mathbf{r}}{\delta s}.$$

When the scalar variable s is the arc-length of the curve traced out by the point whose position vector is **r**, the derivative is frequently denoted by **r′**. Its direction is that of the tangent to the curve at the point considered (Fig. 1, Art. 1).

The derivative is usually itself a function of the scalar variable. Its derivative is called the *second derivative* of **r** with respect to s, and is written

$$\frac{d}{ds}\left(\frac{d\mathbf{r}}{ds}\right) = \frac{d^2\mathbf{r}}{ds^2} = \mathbf{r}'',$$

and so on for derivatives of higher order. If

$$\mathbf{r} = (x,\, y,\, z),$$

then clearly

$$\mathbf{r}' = (x',\, y',\, z')$$

and

$$\mathbf{r}'' = (x'',\, y'',\, z'').$$

If s is a function of another scalar variable t, then, as usual,

$$\frac{d\mathbf{r}}{dt} = \frac{d\mathbf{r}}{ds}\frac{ds}{dt}.$$

The ordinary rules of differentiation hold for sums and products of vectors*. Thus

$$\frac{d}{dt}(\mathbf{r}+\mathbf{s}+\ldots) = \frac{d\mathbf{r}}{dt} + \frac{d\mathbf{s}}{dt} + \ldots,$$

$$\frac{d}{dt}(\mathbf{r}\cdot\mathbf{s}) = \frac{d\mathbf{r}}{dt}\cdot\mathbf{s} + \mathbf{r}\cdot\frac{d\mathbf{s}}{dt},$$

$$\frac{d}{dt}(\mathbf{r}\times\mathbf{s}) = \frac{d\mathbf{r}}{dt}\times\mathbf{s} + \mathbf{r}\times\frac{d\mathbf{s}}{dt}.$$

* *Elem. Vect. Anal.*, Art. 56.

If r is the modulus of $\mathbf{r}$, then $\mathbf{r}^2 = r^2$. Hence on differentiating this formula we have

$$\mathbf{r} \cdot \frac{d\mathbf{r}}{dt} = r \frac{dr}{dt},$$

which is an important result. In particular if $\mathbf{n}$ is a vector of constant length, but variable direction, we have

$$\mathbf{n} \cdot \frac{d\mathbf{n}}{dt} = 0.$$

Thus *a vector of constant length is perpendicular to its derivative.* This property is one of frequent application.

To differentiate a product of several vectors, differentiate each in turn, and take the sum of the products so obtained. For instance

$$\frac{d}{dt}[\mathbf{a}, \mathbf{b}, \mathbf{c}] = \left[\frac{d\mathbf{a}}{dt}, \mathbf{b}, \mathbf{c}\right] + \left[\mathbf{a}, \frac{d\mathbf{b}}{dt}, \mathbf{c}\right] + \left[\mathbf{a}, \mathbf{b}, \frac{d\mathbf{c}}{dt}\right].$$

Suppose next that $\mathbf{r}$ is a function of *several independent variables* $u, v, w, \dots$. Let the first variable increase from u to $u + \delta u$, while the others remain unaltered, and let $\delta\mathbf{r}$ be the corresponding increment in the vector. Then the limiting value of $\delta\mathbf{r}/\delta u$, as δu tends to zero, is called the partial derivative of $\mathbf{r}$ with respect to u, and is written $\dfrac{\partial \mathbf{r}}{\partial u}$. Similarly for partial derivatives with respect to the other variables.

These derivatives, being themselves functions of the same set of variables, may be again differentiated partially, yielding second order partial derivatives. We denote the derivatives of $\dfrac{\partial \mathbf{r}}{\partial u}$ with respect to u and v respectively by

$$\frac{\partial^2 \mathbf{r}}{\partial u^2} \quad \text{and} \quad \frac{\partial^2 \mathbf{r}}{\partial u \partial v},$$

and, as in the scalar calculus,

$$\frac{\partial^2 \mathbf{r}}{\partial u \partial v} = \frac{\partial^2 \mathbf{r}}{\partial v \partial u}.$$

Also, in the notation of differentials, the *total differential* of $\mathbf{r}$ is given by the formula

$$d\mathbf{r} = \frac{\partial \mathbf{r}}{\partial u}\,du + \frac{\partial \mathbf{r}}{\partial v}\,dv + \dots.$$

And, if $\mathbf{n}$ is a vector of constant length, $\mathbf{n}^2 = \text{const.}$, and therefore

$$\mathbf{n} \cdot d\mathbf{n} = 0.$$

Thus a vector of constant length is perpendicular to its differential

In the *geometry of surfaces,* the various quantities are usually functions of two independent variables (or parameters) u, v. Partial derivatives with respect to these are frequently indicated by the use of suffixes 1 and 2 respectively. Thus

$$\mathbf{r}_1 = \frac{\partial \mathbf{r}}{\partial u}, \quad \mathbf{r}_2 = \frac{\partial \mathbf{r}}{\partial v},$$

$$\mathbf{r}_{11} = \frac{\partial^2 \mathbf{r}}{\partial u^2}, \quad \mathbf{r}_{12} = \frac{\partial^2 \mathbf{r}}{\partial u \partial v}, \quad \mathbf{r}_{22} = \frac{\partial^2 \mathbf{r}}{\partial v^2},$$

and so on. The total differential of $\mathbf{r}$ is thus

$$d\mathbf{r} = \mathbf{r}_1 \, du + \mathbf{r}_2 \, dv.$$

SHORT COURSE

In the following pages the Articles marked with an asterisk may be omitted at the first reading.

The student who wishes to take first a short course of what is most essential in the development of the subject should read the following Articles:

1—5, 8, 13—17, 22—43, 46—53, 54 (first part),
56—57, 67—75, 84—86, 91—101.

The reader who is anxious to begin the study of Differential Invariants (Chap. XII) may do so at any stage after Chap. VI.

CHAPTER I

CURVES WITH TORSION

1. Tangent. A *curve* is the locus of a point whose position vector **r** relative to a fixed origin may be expressed as a function of a single variable parameter. Then its Cartesian coordinates x, y, z are also functions of the same parameter. When the curve is not a plane curve it is said to be *skew*, tortuous or twisted. We shall confine our attention to those portions of the curve which are free from singularities of all kinds.

It is usually convenient to choose as the scalar parameter the length s of the arc of the curve measured from a fixed point A on it. Then for points on one side of A the value of s will be positive; for points on the other side, negative. The positive direction along the curve at any point is taken as that corresponding to algebraical increase of s. Thus the position vector **r** of a point on the curve is a function of s, regular within the range considered. Its successive

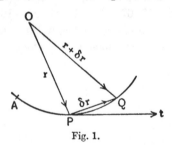

Fig. 1.

derivatives with respect to s will be denoted by **r**′, **r**″, **r**‴, and so on. Let P, Q be the points on the curve whose position vectors are **r**, **r** + δ**r** corresponding to the values s, $s + \delta s$ of the parameter; then δ**r** is the vector PQ. The quotient δ**r**/δs is a vector in the same direction as δ**r**; and in the limit, as δs tends to zero, this direction becomes that of the tangent at P. Moreover the ratio of the lengths of the chord PQ and the arc PQ tends to unity as Q moves up to coincidence with P. Therefore the limiting value of $\delta\mathbf{r}/\delta s$ is a unit vector parallel to the tangent to the curve at P, and in the positive direction. We shall denote this by $\mathbf{t}$ and call it the *unit tangent* at P. Thus

$$\mathbf{t} = \operatorname*{Lt}_{\delta s \to 0} \frac{\delta\mathbf{r}}{\delta s} = \frac{d\mathbf{r}}{ds} = \mathbf{r}' \quad\ldots\ldots\ldots\ldots\ldots(1).$$

The vector equation of the tangent at P may be written down at once. For the position vector $\mathbf{R}$ of a current point on the tangent is given by

$$\mathbf{R} = \mathbf{r} + u\mathbf{t},$$

where u is a variable number, positive or negative. This is the equation of the tangent. If x, y, z are the Cartesian coordinates of P referred to fixed rectangular axes through the same origin, and $\mathbf{i}$, $\mathbf{j}$, $\mathbf{k}$ are unit vectors in the positive directions of these axes,

$$\mathbf{r} = x\mathbf{i} + y\mathbf{j} + z\mathbf{k}$$

and $$\mathbf{t} = \mathbf{r}' = x'\mathbf{i} + y'\mathbf{j} + z'\mathbf{k}.$$

The direction cosines of the tangent are therefore x', y', z'. The *normal plane* at P is the plane through P perpendicular to the tangent. Hence its equation is

$$(\mathbf{R} - \mathbf{r})\cdot\mathbf{t} = 0.$$

Every line through P in this plane is a *normal* to the curve.

2. Principal normal. Curvature. The *curvature* of the curve at any point is the arc-rate of rotation of the tangent. Thus if $\delta\theta$ is the angle between the tangents at P and Q (Fig. 1), $\delta\theta/\delta s$ is the average curvature of the arc PQ; and its limiting value as δs tends to zero is the curvature at the point P. This is sometimes called the first curvature or the *circular curvature*. We shall denote it by κ. Thus

$$\kappa = \operatorname*{Lt}_{\delta s \to 0} \frac{\delta\theta}{\delta s} = \frac{d\theta}{ds} = \theta'.$$

The unit tangent is not a constant vector, for its direction changes from point to point of the curve. Let $\mathbf{t}$ be its value at P and $\mathbf{t} + \delta\mathbf{t}$ at Q. If the vectors BE and BF are respectively equal to these, then $\delta\mathbf{t}$ is the vector EF and $\delta\theta$ the angle EBF. The quotient $\delta\mathbf{t}/\delta s$ is a vector parallel to $\delta\mathbf{t}$, and therefore in the limit as δs tends to zero its direction is perpendicular to the tangent at P. Moreover since *BE* and *BF* are of unit length, the modulus of the limiting

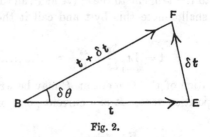

Fig. 2.

value of $\delta t/\delta s$ is the limiting value of $\delta\theta/\delta s$, which is κ. Hence the relation

$$\frac{dt}{ds} = \mathrm{Lt}\,\frac{\delta t}{\delta s} = \kappa n \quad\text{......................}(2),$$

where **n** is a *unit* vector perpendicular to **t**, and in the plane of the tangents at *P* and a consecutive point. This plane, containing two consecutive tangents and therefore three consecutive points at *P*, is called the *plane of curvature* or the *osculating plane* at *P*. If **R** is any point in this plane the vectors **R** − **r**, **t** and **n** are coplanar. Hence the relation

$$[\mathbf{R} - \mathbf{r},\ \mathbf{t},\ \mathbf{n}] = 0,$$

which is the equation of the osculating plane. It may also be expressed

$$[\mathbf{R} - \mathbf{r},\ \mathbf{r}',\ \mathbf{r}''] = 0.$$

The unit vectors **t** and **n** are perpendicular to each other, and their plane is the plane of curvature. The straight line through *P* parallel to **n** is called the *principal normal* at *P*. Its equation is clearly

$$\mathbf{R} = \mathbf{r} + u\mathbf{n},$$

R being a current point on the line. The vector **n** will be called the *unit (principal) normal*. It may be assigned either of the two opposite directions along the principal normal. If we give it that from *P* toward the concave side of the curve it follows from (2) that κ must be regarded as positive, for **t**′ has also this direction. But it is sometimes convenient to take **n** as being directed from the curve toward its convex side. In this case κ is negative, for **t**′

and **n** have then opposite directions*. Writing (2) in the form

$$\mathbf{n} = \frac{\mathbf{r}''}{\kappa} = \frac{1}{\kappa}(x''\mathbf{i} + y''\mathbf{j} + z''\mathbf{k}),$$

and observing that **n** is a unit vector, we see that its direction cosines are $\dfrac{x''}{\kappa}, \dfrac{y'}{\kappa}, \dfrac{z''}{\kappa}$. Also on squaring (2) we find the formula

$$\kappa^2 = \mathbf{r}''^2 = x''^2 + y''^2 + z''^2$$

for determining the magnitude of the curvature, though not its sign.

The *circle of curvature* at P is the circle passing through three points on the curve ultimately coincident at P. Its radius ρ is called the *radius of (circular) curvature*, and its centre C the *centre of curvature*. This circle clearly lies in the osculating plane at P, and its curvature is the same as that of the curve at P, for it has two consecutive tangents in common with the curve. Thus

$$\frac{1}{\rho} = \frac{d\theta}{ds} = \kappa,$$

so that the radius of curvature is the reciprocal of the curvature, and must be regarded as having the same sign as the curvature. The centre of curvature C lies on the principal normal, and the vector PC is equal to $\rho\mathbf{n}$ or $\mathbf{n}/\kappa$. The direction cosines of **n** as found above may now be written $\rho x'', \rho y'', \rho z''$, and the equation (2) given the alternative form

$$\mathbf{n} = \rho\mathbf{t}' = \rho\mathbf{r}''.$$

3. Binormal. Torsion. Among the normals at P to the curve that which is perpendicular to the osculating plane is called the *binormal*. Being perpendicular to both **t** and **n** it is parallel to $\mathbf{t} \times \mathbf{n}$. Denoting this unit vector by **b** we have the trio **t**, **n**, **b** forming a right-handed system of mutually perpendicular unit vectors, and therefore connected by the relations

$$\mathbf{t} \cdot \mathbf{n} = \mathbf{n} \cdot \mathbf{b} = \mathbf{b} \cdot \mathbf{t} = 0$$

and $$\mathbf{t} \times \mathbf{n} = \mathbf{b}, \quad \mathbf{n} \times \mathbf{b} = \mathbf{t}, \quad \mathbf{b} \times \mathbf{t} = \mathbf{n},$$

the cyclic order being preserved in the cross products. We may

---

* This is a departure from the usual practice of regarding κ as essentially positive.

call **b** the *unit binormal.* The positive direction along the binormal is taken as that of **b**, just as the positive direction along the principal normal is that of **n**. The equation of the binormal is

$$\mathbf{R} = \mathbf{r} + u\mathbf{b},$$

or, since $\mathbf{b} = \mathbf{t} \times \mathbf{n} = \rho\mathbf{r}' \times \mathbf{r}''$, we may write it in the form

$$\mathbf{R} = \mathbf{r} + v\mathbf{r}' \times \mathbf{r}''.$$

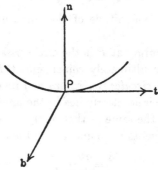

Fig. 3.

The direction cosines of the binormal, being the resolved parts of **b**, are the resolved parts of $\rho\mathbf{r}' \times \mathbf{r}''$, and are therefore

$$\rho\,(y'z'' - z'y''), \quad \rho\,(z'x'' - x'z''), \quad \rho\,(x'y'' - y'x'').$$

Since **b** is a vector of constant length it follows that **b**′ is perpendicular to **b**. Moreover by differentiating the relation $\mathbf{t} \cdot \mathbf{b} = 0$ with respect to s we find

$$\kappa\mathbf{n} \cdot \mathbf{b} + \mathbf{t} \cdot \mathbf{b}' = 0.$$

The first term vanishes because **n** is perpendicular to **b**, and the equation then shows that **b**′ is perpendicular to **t**. But it is also perpendicular to **b**, and must therefore be parallel to **n**. We may then write

$$\frac{d\mathbf{b}}{ds} = -\tau\mathbf{n} \quad \dotfill (3).$$

Just as in equation (2) the scalar κ measures the arc-rate of turning of the unit vector **t**, so here τ measures the arc-rate of turning of the unit vector **b**. This rate of turning of the binormal is called the *torsion* of the curve at the point P. It is of course the rate of rotation of the osculating plane. The negative sign in (3) indicates that the torsion is regarded as positive when the rotation of the binormal as s increases is in the same sense as that of a right-handed screw travelling in the direction of $\mathbf{t}$. It is clear from Fig. 3 that in this case $\mathbf{b}'$ has the opposite direction to $\mathbf{n}$.

The derivative of $\mathbf{n}$ may now be deduced from those of $\mathbf{t}$ and $\mathbf{b}$. For

$$\frac{d\mathbf{n}}{ds} = \frac{d}{ds}\,\mathbf{b} \times \mathbf{t} = -\tau\mathbf{n} \times \mathbf{t} + \mathbf{b} \times (\kappa\mathbf{n})$$

$$= \tau\mathbf{b} - \kappa\mathbf{t} \quad\quad\quad\quad\quad\quad\quad\quad\quad (4).$$

The equations (2), (3) and (4) are the vector equivalents of the **Serret-Frenet formulae.** They will be much used in the following pages, and the reader should commit them to memory. We may gather them together in the form

$$\left.\begin{array}{l} \mathbf{t}' = \kappa\mathbf{n} \\ \mathbf{n}' = \tau\mathbf{b} - \kappa\mathbf{t} \\ \mathbf{b}' = -\tau\mathbf{n} \end{array}\right\} \quad\quad\quad\quad (5).$$

As given by Serret (1851) and Frenet (1852) these were formulae for the derivatives of the direction cosines of the tangent, the principal normal and the binormal.

A formula for the torsion in terms of the derivatives of $\mathbf{r}$ may now be found. For

$$\mathbf{r}' = \mathbf{t}, \quad \mathbf{r}'' = \kappa\mathbf{n},$$

and therefore $\quad\quad \mathbf{r}''' = \kappa'\mathbf{n} + \kappa\,(\tau\mathbf{b} - \kappa\mathbf{t}).$

Forming then the scalar triple product of these three derivatives and neglecting those triple products in the expansion which contain a repeated factor, we have

$$[\mathbf{r}', \mathbf{r}'', \mathbf{r}'''] = [\mathbf{t}, \kappa\mathbf{n}, \kappa'\mathbf{n} + \kappa\,(\tau\mathbf{b} - \kappa\mathbf{t})]$$

$$= \kappa^2\tau\,[\mathbf{t}, \mathbf{n}, \mathbf{b}]$$

$$= \kappa^2\tau.$$

Hence the value of the torsion is given by

$$\tau = \frac{1}{\kappa^2}[\mathbf{r}', \mathbf{r}'', \mathbf{r}'''] \quad\quad\quad\quad (6).$$

An alternative formula, giving the square of the torsion, may be deduced from the expression for $\mathbf{r}'''$ found above. On squaring this and dividing throughout by κ^2 we obtain the result

$$\tau^2 = \frac{1}{\kappa^2}\mathbf{r}'''^2 - \kappa^2 - \left(\frac{\kappa'}{\kappa}\right)^2 \quad \ldots\ldots\ldots\ldots(7).$$

By analogy with the relation that the radius of curvature is equal to the reciprocal of the curvature, it is customary to speak of the reciprocal of the torsion as the *radius of torsion,* and to denote it by σ. Thus $\sigma = 1/\tau$. But there is no circle of torsion or centre of torsion associated with the curve in the same way as the circle and centre of curvature.

Ex. 1. *The circular helix.* This is a curve drawn on the surface of a circular cylinder, cutting the generators at a constant angle β. Let a be the radius of the cylinder, and let its axis be taken as the axis of z. The plane through the axis and the point (x, y, z) on the helix is inclined to the zx plane at an angle θ such that $x = a \cos \theta$ and $y = a \sin \theta$, while $z = a\theta \cot \beta$. The position vector $\mathbf{r}$ of a point on the curve may then be expressed

$$\mathbf{r} = a\,(\cos \theta,\ \sin \theta,\ \theta \cot \beta).$$

Differentiating with respect to s we have

$$\mathbf{t} = \mathbf{r}' = a\,(-\sin \theta,\ \cos \theta,\ \cot \beta)\,\theta'.$$

But this is a unit vector, so that its square is unity, and therefore

$$a^2\theta'^2 = \sin^2 \beta.$$

Thus θ' is constant. To find the curvature we have, on differentiating $\mathbf{t}$,

$$\kappa\mathbf{n} = \mathbf{r}'' = -a\,(\cos \theta,\ \sin \theta,\ 0)\,\theta'^2.$$

Thus the principal normal is the unit vector

$$\mathbf{n} = -(\cos \theta,\ \sin \theta,\ 0),$$

and $$\kappa = a\theta'^2 = \frac{1}{a}\sin^2 \beta.$$

To find the torsion we have

$$\mathbf{r}''' = a\,(\sin \theta,\ -\cos \theta,\ 0)\,\theta'^3,$$

and therefore $$\mathbf{r}'' \times \mathbf{r}''' = a^2\,(0,\ 0,\ 1)\,\theta'^5.$$

Hence $$\kappa^2\tau = [\mathbf{r}',\ \mathbf{r}'',\ \mathbf{r}'''] = a^3 \cot \beta\,\theta'^6.$$

On substituting the values of κ and θ' we find

$$\tau = \frac{1}{a}\sin \beta \cos \beta.$$

Thus the curvature and the torsion are both constant, and therefore their ratio is constant. The principal normal intersects the axis of the cylinder orthogonally; and the tangent and binormal are inclined at constant angles to the fixed direction of the generators.

Ex. 2. For the curve

$$x = a\,(3u - u^3),\quad y = 3au^2,\quad z = a\,(3u + u^3),$$

show that
$$\kappa = \tau = \frac{1}{3a\,(1 + u^2)^2}\,.$$

4. Locus of centre of curvature. Just as the arc-rate of turning of the tangent is called the curvature, and the arc-rate of turning of the binormal the torsion, so the arc-rate of turning of the principal normal is called the *screw curvature*. Its magnitude is the modulus of **n'**. But we have seen that

$$\mathbf{n'} = \tau\mathbf{b} - \kappa\mathbf{t}.$$

Hence the magnitude of the screw curvature is $\sqrt{\kappa^2 + \tau^2}$. This quantity, however, does not play such an important part in the theory of curves as the curvature and torsion.

The centre of curvature at P is the point of intersection of the principal normal at P with that normal at the consecutive point P' which lies in the osculating plane at P. Consecutive principal normals do not in general intersect (cf. Ex. 8 below). It is worth noticing that the tangent to the locus of the centre of curvature lies in the normal plane of the original curve. For the centre of curvature is the point whose position vector **c** is given by

$$\mathbf{c} = \mathbf{r} + \rho\mathbf{n},$$

The tangent to its locus, being parallel to $\dfrac{d\mathbf{c}}{ds}$, is therefore parallel

to
$$\mathbf{t} + \rho'\mathbf{n} + \rho\,(\tau\mathbf{b} - \kappa\mathbf{t}),$$

that is, to
$$\rho'\mathbf{n} + \rho\tau\mathbf{b}.$$

It therefore lies in the normal plane of the original curve, and is inclined to the principal normal **n** at an angle β such that

$$\tan\beta = \frac{\rho\tau}{\rho'} = \frac{\rho}{\rho'\sigma}\,.$$

If the original curve is one of constant curvature, $\rho' = 0$, and the tangent to the locus of C is then parallel to **b**. It will be proved in Art. 6 that the locus of C has then the same constant curvature as the original curve, and that its torsion varies inversely as the torsion of the given curve. (Cf. also Ex. 19 below.)

EXAMPLES I

1. Prove that
$$\mathbf{r}''' = \kappa'\mathbf{n} - \kappa^2\mathbf{t} + \kappa\tau\mathbf{b},$$
and hence that $\quad \mathbf{r}'''' = (\kappa'' - \kappa^3 - \kappa\tau^2)\mathbf{n} - 3\kappa\kappa'\mathbf{t} + (2\kappa'\tau + \tau'\kappa)\mathbf{b}.$

2. Prove the relations
$$\mathbf{r}' \cdot \mathbf{r}'' = 0, \quad \mathbf{r}' \cdot \mathbf{r}''' = -\kappa^2, \quad \mathbf{r}' \cdot \mathbf{r}'''' = -3\kappa\kappa',$$
$$\mathbf{r}'' \cdot \mathbf{r}''' = \kappa\kappa', \quad \mathbf{r}'' \cdot \mathbf{r}'''' = \kappa(\kappa'' - \kappa^3 - \kappa\tau^2),$$
$$\mathbf{r}''' \cdot \mathbf{r}'''' = \kappa'\kappa'' + 2\kappa^3\kappa' + \kappa^2\tau\tau' + \kappa\kappa'\tau^2.$$

3. If the nth derivative of $\mathbf{r}$ with respect to s is given by
$$\mathbf{r}^{(n)} = a_n\mathbf{t} + b_n\mathbf{n} + c_n\mathbf{b},$$
prove the reduction formulae
$$a_{n+1} = a_n' - \kappa b_n,$$
$$b_{n+1} = b_n' + \kappa a_n - \tau c_n,$$
$$c_{n+1} = c_n' + \tau b_n.$$

4. If κ is zero at all points, the curve is a straight line. If τ is zero at all points, the curve is plane. The necessary and sufficient condition that the curve be plane is
$$[\mathbf{r}', \mathbf{r}'', \mathbf{r}'''] \equiv 0.$$

5. Prove that for any curve $\mathbf{t} \cdot \mathbf{b}' = -\kappa\tau.$

6. If the tangent and the binormal at a point of a curve make angles θ, ϕ respectively with a fixed direction, show that
$$\frac{\sin\theta}{\sin\phi}\frac{d\theta}{d\phi} = -\frac{\kappa}{\tau}.$$

7. Coordinates in terms of s. If s is the arc-length measured from a fixed point A on the curve to the current point P, the position vector $\mathbf{r}$ of P is a function of s; and therefore, by Taylor's Theorem,
$$\mathbf{r} = \mathbf{r}_0 + s\mathbf{r}_0' + \frac{s^2}{\lfloor 2}\mathbf{r}_0'' + \frac{s^3}{\lfloor 3}\mathbf{r}_0''' + \ldots\ldots,$$
where the suffix zero indicates that the value of the quantity is to be taken for the point A. If $\mathbf{t}$, $\mathbf{n}$, $\mathbf{b}$ are the unit tangent, principal normal and binormal at A, and κ, τ the curvature and torsion at that point, we have
$$\mathbf{r}_0' = \mathbf{t}, \quad \mathbf{r}_0'' = \kappa\mathbf{n},$$
while the values of $\mathbf{r}_0'''$ and $\mathbf{r}_0''''$ are as given in Ex. 1. Hence the above formula gives
$$\mathbf{r} = \mathbf{r}_0 + s\mathbf{t} + \frac{s^2}{\lfloor 2}\kappa\mathbf{n} + \frac{s^3}{\lfloor 3}(\kappa'\mathbf{n} - \kappa^2\mathbf{t} + \kappa\tau\mathbf{b})$$
$$+ \frac{s^4}{\lfloor 4}\{(\kappa'' - \kappa^3 - \kappa\tau^2)\mathbf{n} - 3\kappa\kappa'\mathbf{t} + (2\kappa'\tau + \kappa\tau')\mathbf{b}\} + \ldots$$

If then A is taken as origin, and the tangent, principal normal and binormal at A as coordinate axes, the coordinates of P are the coefficients of $\mathbf{t}$, $\mathbf{n}$, $\mathbf{b}$ in the above expansion. Thus since $\mathbf{r}_0$ is now zero, we have

$$x = s - \frac{\kappa^2}{6}s^3 - \frac{\kappa\kappa'}{8}s^4 + \ldots\ldots,$$

$$y = \frac{\kappa}{2}s^2 + \frac{\kappa'}{6}s^3 + \frac{1}{24}(\kappa'' - \kappa^3 - \kappa\tau^2)s^4 + \ldots\ldots,$$

$$z = \frac{1}{6}\kappa\tau s^3 + \frac{1}{24}(2\kappa'\tau + \kappa\tau')s^4 + \ldots\ldots$$

From the last equation it follows that, for sufficiently small values of s, z changes sign with s (unless κ or τ is zero). Hence, at an ordinary point of the curve, the curve crosses the osculating plane. On the other hand, for sufficiently small values of s, y does not change sign with s ($\kappa \neq 0$). Thus, in the neighbourhood of an ordinary point, the curve lies on one side of the plane determined by the tangent and binormal. This plane is called the *rectifying plane*.

8. Show that the principal normals at consecutive points do not intersect unless $\tau = 0$.

Let the consecutive points be $\mathbf{r}$, $\mathbf{r} + d\mathbf{r}$ and the unit principal normals $\mathbf{n}$, $\mathbf{n} + d\mathbf{n}$. For intersection of the principal normals the necessary condition is that the three vectors $d\mathbf{r}$, $\mathbf{n}$, $\mathbf{n} + d\mathbf{n}$ be coplanar: that is, that $\mathbf{r}'$, $\mathbf{n}$, $\mathbf{n}'$ be coplanar. This requires

$$[\mathbf{t}, \mathbf{n}, \tau\mathbf{b} - \kappa\mathbf{t}] = 0,$$

that is $\tau[\mathbf{t}, \mathbf{n}, \mathbf{b}] = 0,$

which holds only when τ vanishes.

9. Prove that the shortest distance between the principal normals at consecutive points, distant s apart, is $s\rho/\sqrt{\rho^2 + \sigma^2}$, and that it divides the radius of curvature in the ratio $\rho^2 : \sigma^2$.

10. Prove that
$$\mathbf{b}'' = \tau(\kappa\mathbf{t} - \tau\mathbf{b}) - \tau'\mathbf{n},$$
$$\mathbf{n}'' = \tau'\mathbf{b} - (\kappa^2 + \tau^2)\mathbf{n} - \kappa'\mathbf{t},$$
and find similar expressions for $\mathbf{b}'''$ and $\mathbf{n}'''$.

11. Parameter other than s. If the position vector $\mathbf{r}$ of the current point is a function of any parameter u, and dashes denote differentiations with respect to u, we have
$$\mathbf{r} = \frac{d\mathbf{r}}{ds}\frac{ds}{du} = s'\mathbf{t},$$
$$\mathbf{r}'' = s''\mathbf{t} + \kappa s'^2\mathbf{n},$$
$$\mathbf{r}''' = (s''' - \kappa^2 s'^3)\mathbf{t} + s'(3\kappa s'' + \kappa's')\mathbf{n} + \kappa\tau s'^3\mathbf{b}.$$

2—2

Hence prove that

$$\mathbf{b} = \mathbf{r}' \times \mathbf{r}''/\kappa s'^3,$$
$$\mathbf{n} = (s'\mathbf{r}'' - s''\mathbf{r}')/\kappa s'^3,$$
$$\kappa^2 = (\mathbf{r}''^2 - s''^2)/s'^4,$$
$$\tau = [\mathbf{r}', \mathbf{r}'', \mathbf{r}''']/\kappa^2 s'^6.$$

12. For the curve

$$x = 4a\cos^5 u, \quad y = 4a\sin^3 u, \quad z = 3c\cos 2u.$$

prove that

$$\mathbf{n} = (\sin u, \cos u, 0),$$

and

$$\kappa = \frac{a}{6(a^2 + c^2)\sin 2u}.$$

13. Find the curvature and torsion of the curve

$$x = a(u - \sin u), \quad y = a(1 - \cos u), \quad z = bu.$$

14. Find the curvature, the centre of curvature, and the torsion of the curve

$$x = a\cos u, \quad y = a\sin u, \quad z = a\cos 2u.$$

15. If the plane of curvature at every point of a curve passes through a fixed point, show that the curve is plane ($\tau = 0$).

16. If $\mathbf{m}_1$, $\mathbf{m}_2$, $\mathbf{m}_3$ are the moments about the origin of unit vectors $\mathbf{t}$, $\mathbf{n}$, $\mathbf{b}$ localised in the tangent, principal normal and binormal, and dashes denote differentiations with respect to s, show that

$$\mathbf{m}_1' = \kappa\mathbf{m}_2, \quad \mathbf{m}_2' = \mathbf{b} - \kappa\mathbf{m}_1 + \tau\mathbf{m}_3, \quad \mathbf{m}_3' = -\mathbf{n} - \tau\mathbf{m}_2.$$

If $\mathbf{r}$ is the current point, we have

$$\mathbf{m}_1 = \mathbf{r} \times \mathbf{t}, \quad \mathbf{m}_2 = \mathbf{r} \times \mathbf{n}, \quad \mathbf{m}_3 = \mathbf{r} \times \mathbf{b}.$$

Therefore

$$\mathbf{m}_1' = \mathbf{t} \times \mathbf{t} + \mathbf{r} \times (\kappa\mathbf{n}) = \kappa\mathbf{m}_2,$$

and similarly for the others.

17. Prove that the position vector of the current point on a curve satisfies the differential equation

$$\frac{d}{ds}\left\{\sigma\frac{d}{ds}\left(\rho\frac{d^2\mathbf{r}}{ds^2}\right)\right\} + \frac{d}{ds}\left(\frac{\sigma}{\rho}\frac{d\mathbf{r}}{ds}\right) + \frac{\rho}{\sigma}\frac{d^2\mathbf{r}}{ds^2} = 0.$$

(Use the Serret-Frenet formulae.)

18. If s_1 is the arc-length of the locus of the centre of curvature, show that

$$\frac{ds_1}{ds} = \frac{1}{\kappa^2}\sqrt{\kappa^2\tau^2 + \kappa'^2} = \sqrt{\left(\frac{\rho}{\sigma}\right)^2 + \rho'^2}.$$

19. In the case of a *curve of constant curvature* find the curvature and torsion of the locus of its centre of curvature C.

The position vector of C is equal to

$$\mathbf{c} = \mathbf{r} + \rho\mathbf{n}.$$

Hence, since ρ is constant,

$$d\mathbf{c} = \{\mathbf{t} + \rho(\tau\mathbf{b} - \kappa\mathbf{t})\}\,ds = \frac{\tau}{\kappa}\mathbf{b}\,ds.$$

Let the suffix unity distinguish quantities belonging to the locus of C. Then $d\mathbf{c} = \mathbf{t}_1 \, ds_1$. We may take the positive direction along this locus so that $\mathbf{t}_1 = \mathbf{b}$. Then it follows that

$$\frac{ds_1}{ds} = \frac{\tau}{\kappa}.$$

Next differentiating the relation $\mathbf{t}_1 = \mathbf{b}$ we obtain

$$\kappa_1 \mathbf{n}_1 = -\tau \mathbf{n} \frac{ds}{ds_1} = -\kappa \mathbf{n}.$$

Therefore the two principal normals are parallel. We may choose

$$\mathbf{n}_1 = -\mathbf{n},$$

and therefore　　　　　　　　　$\kappa_1 = \kappa.$

Thus the locus of C has the same constant curvature as the given curve. The unit binormal $\mathbf{b}_1$ is now fixed: for

$$\mathbf{b}_1 = \mathbf{t}_1 \times \mathbf{n}_1 = \mathbf{b} \times (-\mathbf{n}) = \mathbf{t}.$$

Differentiating this result we obtain

$$-\tau_1 \mathbf{n}_1 = \kappa \mathbf{n} \frac{ds}{ds_1} = \frac{\kappa^2}{\tau} \mathbf{n},$$

and therefore　　　　　　　　　$\tau_1 = \kappa^2/\tau.$

20. Prove that, for any curve,

$$[\mathbf{t}', \mathbf{t}'', \mathbf{t}'''] = [\mathbf{r}'', \mathbf{r}''', \mathbf{r}''''] = \kappa^3 (\kappa\tau' - \kappa'\tau) = \kappa^5 \frac{d}{ds}\left(\frac{\tau}{\kappa}\right),$$

and also that　　　　$[\mathbf{b}', \mathbf{b}'', \mathbf{b}'''] = \tau^3 (\kappa'\tau - \kappa\tau') = \tau^5 \frac{d}{ds}\left(\frac{\kappa}{\tau}\right).$

*5. Spherical curvature. The sphere of closest contact with the curve at P is that which passes through four points on the

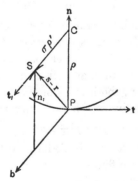

Fig. 4.

curve ultimately coincident with P. This is called the osculating sphere or the *sphere of curvature* at P. Its centre S and radius R

are called the centre and radius of spherical curvature. The centre of a sphere through P and an adjacent point Q on the curve (Fig. 1) lies on the plane which is the perpendicular bisector of PQ; and the limiting position of this plane is the normal plane at P. Thus the centre of spherical curvature is the limiting position of the intersection of three normal planes at adjacent points. Now the normal plane at the point $\mathbf{r}$ is

$$(\mathbf{s} - \mathbf{r})\cdot\mathbf{t} = 0 \quad\dotfill\text{(i)},$$

$\mathbf{s}$ being the current point on the plane. The limiting position of the line of intersection of this plane and an adjacent normal plane is determined by (i) and the equation obtained by differentiating it with respect to the arc-length s, viz. (cf. Arts. 15, 19)

$$\kappa(\mathbf{s} - \mathbf{r})\cdot\mathbf{n} - 1 = 0$$

or its equivalent

$$(\mathbf{s} - \mathbf{r})\cdot\mathbf{n} = \rho \quad\dotfill\text{(ii)}.$$

The limiting position of the point of intersection of three adjacent normal planes is then found from (i), (ii) and the equation obtained by differentiating (ii), viz.

$$(\mathbf{s} - \mathbf{r})\cdot(\tau\mathbf{b} - \kappa\mathbf{t}) = \rho'$$

which, in virtue of (i), is equivalent to

$$(\mathbf{s} - \mathbf{r})\cdot\mathbf{b} = \sigma\rho' \quad\dotfill\text{(iii)}.$$

The vector $\mathbf{s} - \mathbf{r}$ satisfying (i), (ii) and (iii) is clearly

$$\mathbf{s} - \mathbf{r} = \rho\mathbf{n} + \sigma\rho'\mathbf{b} \quad\dotfill\text{(8)},$$

and this equation determines the position vector $\mathbf{s}$ of the centre of spherical curvature. Now $\rho\mathbf{n}$ is the vector PC, and therefore $\sigma\rho'\mathbf{b}$ is the vector CS. Thus the centre of spherical curvature is on the axis of the circle of curvature, at a distance $\sigma\rho'$ from the centre of curvature. On squaring both sides of the last equation we have for determining the radius of spherical curvature

$$R^2 = \rho^2 + \sigma^2\rho'^2 \quad\dotfill\text{(9)}.$$

Another formula for R^2 may be deduced as follows. On squaring the expansion for $\mathbf{r}'''$ we find

$$\mathbf{r}'''^2 = \kappa'^2 + \kappa^4 + \kappa^2\tau^2$$
$$= \kappa^4(1 + \tau^2R^2), \text{ by (9)}.$$

Hence the formula
$$R^2 = \rho^4\sigma^2 r'''^2 - \sigma^2,$$
which is, however, not so important as (9).

For a curve of constant curvature, $\rho' = 0$, and the centre of spherical curvature coincides with the centre of circular curvature (cf. Art. 4).

6. Locus of centre of spherical curvature.

The position vector s of the centre of spherical curvature has been shown to be
$$s = r + \rho n + \sigma\rho' b.$$
Hence, for a small displacement ds of the current point P along the original curve, the displacement of S is
$$ds = \{t + \rho'n + \rho(\tau b - \kappa t) + \sigma'\rho'b + \sigma\rho''b - \rho'n\}\,ds$$
$$= ds\left(\frac{\rho}{\sigma} + \sigma'\rho' + \sigma\rho''\right)b.$$

Thus the tangent to the locus of S is parallel to b (Fig. 4). We may measure the arc-length s_1 of the locus of S in that direction which makes its unit tangent t_1 have the same direction as b. Thus
$$t_1 = b,$$
and, since $ds = t_1 ds_1$, it follows that
$$\frac{ds_1}{ds} = \frac{\rho}{\sigma} + \rho'\sigma' + \sigma\rho''$$
$$= \frac{\rho}{\sigma} + \frac{d}{ds}(\sigma\rho').$$

To find the curvature κ_1 of the locus of S differentiate the equation $t_1 = b$, thus obtaining
$$\kappa_1 n_1 = \frac{db}{ds}\frac{ds}{ds_1} = -\tau n\frac{ds}{ds_1}.$$

Thus the principal normal to the locus of S is parallel to the principal normal of the original curve. We may choose the direction of n_1 as opposite to that of n. Thus
$$n_1 = -n.$$

The unit binormal b_1 of the locus of S is then
$$b_1 = t_1 \times n_1 = b \times (-n) = t,$$
and is thus equal to the unit tangent of the original curve.

The curvature κ_1 as found above is thus equal to

$$\kappa_1 = \tau \frac{ds}{ds_1}.$$

The torsion τ_1 is obtained by differentiating $\mathbf{b}_1 = \mathbf{t}$. Thus

$$-\tau_1 \mathbf{n}_1 = \frac{d\mathbf{t}}{ds}\frac{ds}{ds_1} = \kappa\mathbf{n}\frac{ds}{ds_1},$$

so that

$$\tau_1 = \kappa\frac{ds}{ds_1}.$$

From the last two results it follows that

$$\kappa\kappa_1 = \tau\tau_1,$$

so that the product of the curvatures of the two curves is equal to the product of their torsions. The binormal of each curve is parallel to the tangent to the other, and their principal normals are parallel but in opposite directions (Fig. 4).

If the original curve is one of *constant curvature*, $\rho' = 0$, and S coincides with the centre of circular curvature. Then

$$\frac{ds_1}{ds} = \frac{\rho}{\sigma} = \frac{\tau}{\kappa},$$

and

$$\kappa_1 = \kappa.$$

Thus the locus of the two centres of curvature has the same (constant) curvature as the original curve   Also

$$\tau_1 = \kappa^2/\tau,$$

so that the product of the torsions of the two curves is equal to the square of their common curvature. The circular helix is a curve of constant curvature.

Ex. 1. If ψ is such that $d\psi = \tau ds$, show that

$$\rho_1 = \frac{1}{\kappa_1} = \rho + \frac{d^2\rho}{d\psi^2},$$

$$\sigma_1 = \frac{1}{\tau_1} = \frac{\rho}{\sigma}\left(\rho + \frac{d^2\rho}{d\psi^2}\right),$$

$$R^2 = \rho^2 + \left(\frac{d\rho}{d\psi}\right)^2,$$

$$R\frac{dR}{d\psi} = \rho + \frac{d^2\rho}{d\psi^2} = \rho_1.$$

Ex. 2. Prove that

$$R\frac{dR}{ds} = \sqrt{R^2 - \rho^2}\,\frac{ds_1}{ds}.$$

Ex. 3. Prove that, for curves drawn on the surface of a sphere,

$$\frac{\rho}{\sigma} + \frac{d}{ds}(\sigma\rho') = 0;$$

that is

$$\rho + \frac{d^2\rho}{d\psi^2} = 0,$$

where $d\psi = \tau\, ds$.

Ex. 4. If the radius of spherical curvature is constant, prove that the curve either lies on the surface of a sphere or else has constant curvature.

Ex. 5. The shortest distance between consecutive radii of spherical curvature divides the radius in the ratio

$$\sigma^2 : \rho^2 \left(\frac{dR}{d\rho}\right)^2.$$

Ex. 6. Show that the radius of spherical curvature of a circular helix is equal to the radius of circular curvature.

7. Theorem. *A curve is uniquely determined, except as to position in space, when its curvature and torsion are given functions of its arc-length s.*

Consider two curves having equal curvatures κ and equal torsions τ for the same value of s. Let $\mathbf{t}, \mathbf{n}, \mathbf{b}$ refer to one curve and $\mathbf{t}_1, \mathbf{n}_1, \mathbf{b}_1$ to the other. Then at points on the curve determined by the same value of s we have

$$\frac{d}{ds}(\mathbf{t}\cdot\mathbf{t}_1) = \mathbf{t}\cdot(\kappa\mathbf{n}_1) + \kappa\mathbf{n}\cdot\mathbf{t}_1,$$

$$\frac{d}{ds}(\mathbf{n}\cdot\mathbf{n}_1) = \mathbf{n}\cdot(\tau\mathbf{b}_1 - \kappa\mathbf{t}_1) + (\tau\mathbf{b} - \kappa\mathbf{t})\cdot\mathbf{n}_1,$$

$$\frac{d}{ds}(\mathbf{b}\cdot\mathbf{b}_1) = \mathbf{b}\cdot(-\tau\mathbf{n}_1) + (-\tau\mathbf{n})\cdot\mathbf{b}_1.$$

Now the sum of the second members of these equations is zero.

Hence

$$\frac{d}{ds}(\mathbf{t}\cdot\mathbf{t}_1 + \mathbf{n}\cdot\mathbf{n}_1 + \mathbf{b}\cdot\mathbf{b}_1) = 0,$$

and therefore

$$\mathbf{t}\cdot\mathbf{t}_1 + \mathbf{n}\cdot\mathbf{n}_1 + \mathbf{b}\cdot\mathbf{b}_1 = \text{const.}$$

Suppose now that the two curves are placed so that their initial points, from which s is measured, coincide, and are then turned (without deformation) till their principal planes at the initial point also coincide. Then, at that point, $\mathbf{t} = \mathbf{t}_1$, $\mathbf{n} = \mathbf{n}_1$, $\mathbf{b} = \mathbf{b}_1$, and the value of the constant in the last equation is 3. Thus

$$\mathbf{t}\cdot\mathbf{t}_1 + \mathbf{n}\cdot\mathbf{n}_1 + \mathbf{b}\cdot\mathbf{b}_1 = 3.$$

But the sum of the cosines of three angles can be equal to 3 only when each of the angles vanishes, or is an integral multiple of 2π. This requires that, at all pairs of corresponding points,

$$t = t_1, \quad n = n_1, \quad b = b_1,$$

so that the principal planes of the two curves are parallel. Moreover, the relation $t = t_1$ may be written

$$\frac{d}{ds}(r - r_1) = 0,$$

so that $\qquad r - r_1 =$ a const. vector.

But this difference vanishes at the initial point; and therefore it vanishes throughout. Thus $r = r_1$ at all corresponding points, and the two curves coincide.

In making the initial points and the principal planes there coincident, we altered only the position and orientation of the curves in space; and the theorem has thus been proved. When a curve is specified by equations giving the curvature and torsion as functions of s

$$\kappa = f(s), \quad \tau = F(s),$$

these are called the *intrinsic equations* of the curve.

8. Helices. A curve traced on the surface of a cylinder, and cutting the generators at a constant angle, is called a *helix*. Thus the tangent to a helix is inclined at a constant angle to a fixed direction. If then t is the unit tangent to the helix, and a a constant vector parallel to the generators of the cylinder, we have

$$t \cdot a = \text{const.}$$

and therefore, on differentiation with respect to s,

$$\kappa\, n \cdot a = 0.$$

Thus, since the curvature of the helix does not vanish, *the principal normal is everywhere perpendicular to the generators.* Hence the fixed direction of the generators is parallel to the plane of t and b; and since it makes a constant angle with t, it also makes a constant angle with b.

An important property of all helices is that *the curvature and torsion are in a constant ratio.* To prove this we differentiate the relation $n \cdot a = 0$, obtaining

$$(\tau b - \kappa t) \cdot a = 0.$$

Thus **a** is perpendicular to the vector $\tau\mathbf{b} - \kappa\mathbf{t}$. But **a** is parallel to the plane of **t** and **b**, and must therefore be parallel to the vector $\tau\mathbf{t} + \kappa\mathbf{b}$, which is inclined to **t** at an angle $\tan^{-1}\kappa/\tau$. But this angle is constant. Therefore the curvature and torsion are in a constant ratio.

Conversely we may prove that *a curve whose curvature and torsion are in a constant ratio is a helix*. Let $\tau = c\kappa$ where c is constant. Then since

$$\mathbf{t}' = \kappa\mathbf{n},$$

and $$\mathbf{b}' = -\tau\mathbf{n} = -c\kappa\mathbf{n},$$

it follows that $$\frac{d}{ds}(\mathbf{b} + c\mathbf{t}) = 0,$$

and therefore $$\mathbf{b} + c\mathbf{t} = \mathbf{a},$$

where **a** is a constant vector. Forming the scalar product of each side with **t** we have

$$\mathbf{t} \cdot \mathbf{a} = c.$$

Thus **t** is inclined at a constant angle to the fixed direction of **a**, and the curve is therefore a helix.

Finally we may show that the curvature and the torsion of a helix are in a constant ratio to the curvature κ_0 of the plane section of the cylinder perpendicular to the generators. Take the z-axis

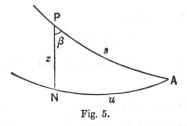

Fig. 5.

parallel to the generators, and let s be measured from the intersection A of the curve with the xy plane. Let u be the arc-length of the normal section of the cylinder by the xy plane, measured from the same point A up to the generator through the current point (x, y, z). Then, if β is the constant angle at which the curve cuts the generators, we have

$$u = s \sin\beta,$$

and therefore $$u' = \sin\beta.$$

The coordinates x, y are functions of u, while $z = s \cos \beta$. Hence for the current point on the helix we have

$$\mathbf{r} = (x,\ y,\ s \cos \beta),$$

so that
$$\mathbf{r}' = \left(\frac{dx}{du} \sin \beta,\ \frac{dy}{du} \sin \beta,\ \cos \beta\right),$$

and
$$\mathbf{r}'' = \left(\frac{d^2x}{du^2} \sin^2 \beta,\ \frac{d^2y}{du^2} \sin^2 \beta,\ 0\right).$$

Hence the curvature of the helix is given by

$$\kappa^2 = \mathbf{r}''^2 = \left\{\left(\frac{d^2x}{du^2}\right)^2 + \left(\frac{d^2y}{du^2}\right)^2\right\} \sin^4 \beta = \kappa_0{}^2 \sin^4 \beta,$$

so that
$$\kappa = \kappa_0 \sin^2 \beta.$$

For the torsion, we have already proved that

$$\beta = \tan^{-1} \kappa/\tau,$$

so that
$$\tau = \kappa \cot \beta = \kappa_0 \sin \beta \cos \beta.$$

From these results it is clear that *the only curve whose curvature and torsion are both constant is the circular helix*. For such a curve must be a helix, since the ratio of its curvature to its torsion is also constant. And since κ is constant it then follows that κ_0 is constant, so that the cylinder on which the helix is drawn is a circular cylinder.

Ex. Show that, for any curve,

$$[\mathbf{r}'',\ \mathbf{r}''',\ \mathbf{r}''''] = \kappa^5 \frac{d}{ds}\left(\frac{\tau}{\kappa}\right).$$

This expression therefore vanishes for a helix: and conversely, if it vanishes, the curve is a helix.

9. Spherical indicatrix. The locus of a point, whose position vector is equal to the unit tangent $\mathbf{t}$ of a given curve, is called the *spherical indicatrix of the tangent* to the curve. Such a locus lies on the surface of a unit sphere; hence the name. Let the suffix unity be used to distinguish quantities belonging to this locus. Then
$$\mathbf{r}_1 = \mathbf{t},$$

and therefore
$$\mathbf{t}_1 = \frac{d\mathbf{r}_1}{ds_1} = \frac{d\mathbf{t}}{ds} \frac{ds}{ds_1} = \kappa \mathbf{n} \frac{ds}{ds_1},$$

showing that the tangent to the spherical indicatrix is parallel to the principal normal of the given curve. We may measure s_1 so that

$$t_1 = n,$$

and therefore

$$\frac{ds_1}{ds} = \kappa.$$

For the curvature κ_1 of the indicatrix, on differentiating the relation $t_1 = n$, we find the formula

$$\kappa_1 n_1 = \frac{dn}{ds}\frac{ds}{ds_1} = \frac{1}{\kappa}(\tau b - \kappa t).$$

Squaring both sides we obtain the result

$$\kappa_1{}^2 = (\kappa^2 + \tau^2)/\kappa^2,$$

so that the curvature of the indicatrix is the ratio of the screw curvature to the circular curvature of the curve. The unit binormal of the indicatrix is

$$b_1 = t_1 \times n_1 = \frac{\tau t + \kappa b}{\kappa\kappa_1}.$$

The torsion could be obtained by differentiating this equation; but the result follows more easily from the equation [cf. Examples I, (11)]

$$\kappa_1{}^2\tau_1\left(\frac{ds_1}{ds}\right)^6 = [r_1', r_1'', r_1'''] = [t', t'', t''']$$
$$= \kappa^3(\kappa\tau' - \kappa'\tau),$$

which reduces to

$$\tau_1 = \frac{(\kappa\tau' - \kappa'\tau)}{\kappa(\kappa^2 + \tau^2)}.$$

Similarly the *spherical indicatrix of the binormal* of the given curve is the locus of a point whose position vector is b. Using the suffix unity to distinguish quantities belonging to this locus, we have

$$r_1 = b,$$

and therefore

$$t_1 = \frac{db}{ds}\frac{ds}{ds_1} = -\tau n \frac{ds}{ds_1}.$$

We may measure s_1 so that

$$t_1 = -n,$$

and therefore

$$\frac{ds_1}{ds} = \tau.$$

To find the curvature differentiate the equation $t_1 = -n$. Then

$$\kappa_1 n_1 = \frac{d}{ds}(-n)\frac{ds}{ds_1} = \frac{1}{\tau}(\kappa t - \tau b),$$

giving the direction of the principal normal. On squaring this result
we have

$$\kappa_1{}^2 = (\kappa^2 + \tau^2)/\tau^2.$$

Thus the curvature of the indicatrix is the ratio of the screw
curvature to the torsion of the given curve. The unit binormal is

$$\mathbf{b}_1 = \mathbf{t}_1 \times \mathbf{n}_1 = \frac{\tau \mathbf{t} + \kappa \mathbf{b}}{\tau \kappa_1},$$

and the torsion, found as in the previous case, is equal to

$$\tau_1 = \frac{\tau \kappa' - \kappa \tau'}{\tau\,(\kappa^2 + \tau^2)}.$$

Ex. 1. Find the torsions of the spherical indicatrices from the formula

$$R^2 = \rho_1{}^2 + \sigma_1{}^2 \rho_1{}'^2,$$

where $R=1$ and $\rho_1 = 1/\kappa_1$ is known.

Ex. 2. Examine the spherical indicatrix of the principal normal of a given
curve.

10. Involutes. When the tangents to a curve C are normals
to another curve C_1, the latter is called an *involute* of the former,
and C is called an *evolute* of C_1. An involute may be generated

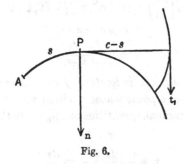

Fig. 6.

mechanically in the following manner. Let one end of an inextensible string be fixed to a point of the curve C, and let the string
be kept taut while it is wrapped round the curve on its convex
side. Then any particle of the string describes an involute of C,
since at each instant the free part of the string is a tangent to
the curve C, while the direction of motion of the particle is at
right angles to this tangent.

From the above definition it follows that the point $\mathbf{r}_1$ of the

involute which lies on the tangent at the point $\mathbf{r}$ of the curve C is given by
$$\mathbf{r}_1 = \mathbf{r} + u\mathbf{t},$$
where u is to be determined. Let ds_1 be the arc-length of the involute corresponding to the element ds of the curve C. Then the unit tangent to C_1 is
$$\mathbf{t}_1 = \frac{d\mathbf{r}_1}{ds}\frac{ds}{ds_1} = \{(1 + u')\mathbf{t} + u\kappa\mathbf{n}\}\frac{ds}{ds_1}.$$
To satisfy the condition for an involute, this vector must be perpendicular to $\mathbf{t}$. Hence
$$1 + u' = 0,$$
so that
$$u = c - s,$$
where c is an arbitrary constant. Thus the current point on the involute is
$$\mathbf{r}_1 = \mathbf{r} + (c - s)\,\mathbf{t},$$
and the unit tangent there is
$$\mathbf{t}_1 = (c - s)\,\kappa\,\frac{ds}{ds_1}\,\mathbf{n}.$$
Hence the tangent to the involute is parallel to the principal normal to the given curve. We may take the positive direction along the involute so that
$$\mathbf{t}_1 = \mathbf{n},$$
and therefore
$$\frac{ds_1}{ds} = (c - s)\,\kappa.$$

To find the curvature κ_1 of the involute we differentiate the relation $\mathbf{t}_1 = \mathbf{n}$, thus obtaining
$$\kappa_1 \mathbf{n}_1 = \frac{\tau\mathbf{b} - \kappa\mathbf{t}}{\kappa(c - s)}.$$
Therefore, on squaring both sides, we have
$$\kappa_1^2 = \frac{\kappa^2 + \tau^2}{\kappa^2 (c - s)^2}.$$
The unit principal normal to the involute is
$$\mathbf{n}_1 = \frac{\tau\mathbf{b} - \kappa\mathbf{t}}{\kappa\kappa_1 (c - s)},$$
and the unit binormal
$$\mathbf{b}_1 = \mathbf{t}_1 \times \mathbf{n}_1 = \frac{\kappa\mathbf{b} + \tau\mathbf{t}}{\kappa\kappa_1 (c - s)}.$$

Since the constant c is arbitrary, there is a single infinitude of involutes to a given curve; and the tangents at corresponding points of two different involutes are parallel and at a constant distance apart.

Ex. 1. Show that the torsion of an involute has the value

$$\frac{\kappa\tau' - \kappa'\tau}{\kappa\,(\kappa^2 + \tau^2)\,(c - s)}.$$

Ex. 2. Prove that the involutes of a circular helix are plane curves, whose planes are normal to the axis of the cylinder, and that they are also involutes of the circular sections of the cylinder.

*11. Evolutes. The converse problem to that just solved is the problem of finding the *evolutes* of a given curve C. Let $\mathbf{r}_1$ be the point on the evolute C_1 corresponding to the point $\mathbf{r}$ on C. Then, since the tangents to C_1 are normals to C, the point $\mathbf{r}_1$ lies in the normal plane to the given curve at $\mathbf{r}$. Hence

$$\mathbf{r}_1 = \mathbf{r} + u\mathbf{n} + v\mathbf{b},$$

where u, v are to be determined. The tangent to the evolute at $\mathbf{r}_1$ is parallel to $d\mathbf{r}_1/ds$, that is, to

$$(1 - u\kappa)\mathbf{t} + (u' - v\tau)\mathbf{n} + (u\tau + v')\mathbf{b}.$$

Hence, in order that it may be parallel to $u\mathbf{n} + v\mathbf{b}$ we must have

$$1 - u\kappa = 0,$$

and

$$\frac{u' - v\tau}{u} = \frac{u\tau + v'}{v}.$$

The first of these gives $u = \dfrac{1}{\kappa} = \rho$, and from the second it follows that

$$\tau = \frac{v\rho' - \rho v'}{v^2 + \rho^2}.$$

Integrating with respect to s and writing $\psi = \displaystyle\int_0^s \tau\,ds$, we have

$$\psi + c = \tan^{-1}\left(-\frac{v}{\rho}\right),$$

so that $\qquad\qquad v = -\rho\tan(\psi + c).$

The point $\mathbf{r}_1$ on the evolute is therefore given by

$$\mathbf{r}_1 = \mathbf{r} + \rho\,\{\mathbf{n} - \tan(\psi + c)\,\mathbf{b}\}.$$

It therefore lies on the axis of the circle of curvature of the given curve, at a distance $-\rho\tan(\psi + c)$ from the centre of curvature.

The tangent to the evolute, being the line joining the points **r** and $\mathbf{r}_1$, is in the normal plane of the given curve C, and is inclined to the principal normal **n** at an angle $(\psi + c)$.

Let the suffix unity distinguish quantities referring to the evolute. Then on differentiating the last equation, remembering that $d\psi/ds = \tau$, we find

$$\mathbf{t}_1 \frac{ds_1}{ds} = \frac{d\mathbf{r}_1}{ds} = \{\rho' + \rho\tau \tan(\psi + c)\}\{\mathbf{n} - \tan(\psi + c)\mathbf{b}\}.$$

Thus the unit tangent to the evolute is

$$\mathbf{t}_1 = \cos(\psi + c)\mathbf{n} - \sin(\psi + c)\mathbf{b}$$

and therefore
$$\frac{ds_1}{ds} = \frac{\kappa\tau \sin(\psi + c) - \kappa' \cos(\psi + c)}{\kappa^2 \cos^2(\psi + c)}.$$

The curvature of the evolute is obtained by differentiating the vector $\mathbf{t}_1$. Thus

$$\kappa_1\mathbf{n}_1 \frac{ds_1}{ds} = \frac{d\mathbf{t}_1}{ds} = -\kappa \cos(\psi + c)\mathbf{t}.$$

The principal normal to the evolute is thus parallel to the tangent to the curve C. We may take

$$\mathbf{n}_1 = -\mathbf{t},$$

and therefore
$$\kappa_1 = \kappa \cos(\psi + c)\frac{ds}{ds_1}$$

$$= \frac{\kappa^3 \cos^3(\psi + c)}{\kappa\tau \sin(\psi + c) - \kappa' \cos(\psi + c)}.$$

The unit binormal to the evolute is

$$\mathbf{b}_1 = \mathbf{t}_1 \times \mathbf{n}_1 = \cos(\psi + c)\mathbf{b} + \sin(\psi + c)\mathbf{n}.$$

The torsion is found by differentiating this. Thus

$$-\tau_1\mathbf{n}_1 \frac{ds_1}{ds} = -\kappa \sin(\psi + c)\mathbf{t}$$

and therefore

$$\tau_1 = -\kappa \sin(\psi + c)\frac{ds}{ds_1}$$

$$= -\frac{\kappa^3 \sin(\psi + c)\cos^2(\psi + c)}{\kappa\tau \sin(\psi + c) - \kappa' \cos(\psi + c)}.$$

Thus the ratio of the torsion of the evolute to its curvature is $-\tan(\psi + c)$.

Since the constant c is arbitrary there is a single infinitude of

w.   **3**

evolutes. The tangents to two different evolutes, corresponding to the values c_1 and c_2, drawn from the same point of the given curve, are inclined to each other at a constant angle $c_1 - c_2$.

Ex. 1. The locus of the centre of curvature is an evolute only when the curve is plane.

Ex. 2. A plane curve has only a single evolute in its own plane, the locus of the centre of curvature. All other evolutes are helices traced on the right cylinder whose base is the plane evolute.

*12. Bertrand curves. Saint-Venant proposed and Bertrand solved the problem of finding the curves whose principal normals are also the principal normals of another curve. A pair of curves, C and C_1, having their principal normals in common, are said to be conjugate or associate *Bertrand curves*. We may take their principal normals in the same sense, so that

$$\mathbf{n}_1 = \mathbf{n}.$$

The point $\mathbf{r}_1$ on C_1 corresponding to the point $\mathbf{r}$ on C is then given by

$$\mathbf{r}_1 = \mathbf{r} + a\mathbf{n} \quad\ldots\ldots\ldots\ldots\ldots\ldots(i),$$

where it is easily seen that a *is constant*. For the tangent to C_1 is parallel to $d\mathbf{r}_1/ds$, and therefore to

$$\mathbf{t} + a'\mathbf{n} + a\,(\tau\mathbf{b} - \kappa\mathbf{t}).$$

This must be perpendicular to $\mathbf{n}$, so that a' is zero and therefore a constant. Further, if symbols with the suffix unity refer to the curve C_1, we have

$$\frac{d}{ds}\,(\mathbf{t}\cdot\mathbf{t}_1) = \kappa\mathbf{n}\cdot\mathbf{t}_1 + \mathbf{t}\cdot(\kappa_1\mathbf{n})\,\frac{ds_1}{ds} = 0,$$

showing that $\qquad\qquad \mathbf{t}\cdot\mathbf{t}_1 = \text{const.}$

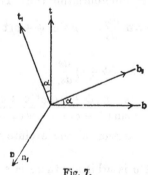

Fig. 7.

Thus *the tangents to the two curves are inclined at a constant angle.*
But the principal normals coincide, and therefore the binormals of
the two curves are inclined at the same constant angle. Let α be
the inclination of $\mathbf{b}_1$ to $\mathbf{b}$ measured from $\mathbf{b}$ toward $\mathbf{t}$. Then α is
constant.

On differentiating the above expression for $\mathbf{r}_1$ we have

$$\mathbf{t}_1 \frac{ds_1}{ds} = (1 - a\kappa)\,\mathbf{t} + a\tau\mathbf{b} \quad\dotfill\text{(ii)}.$$

Then, forming the scalar product of each side with $\mathbf{b}_1$, we obtain

$$0 = (1 - a\kappa)\sin\alpha + a\tau\cos\alpha.$$

Thus *there is a linear relation with constant coefficients between the
curvature and torsion of C;*

$$\tau = \left(\kappa - \frac{1}{a}\right)\tan\alpha.$$

Moreover it is obvious from the diagram that

$$\mathbf{t}_1 = \mathbf{t}\cos\alpha - \mathbf{b}\sin\alpha.$$

On comparing this with (ii) we see that

$$\left.\begin{aligned}
\cos\alpha &= (1 - a\kappa)\frac{ds}{ds_1}\\
\sin\alpha &= -\,a\tau\frac{ds}{ds_1}
\end{aligned}\right\} \quad\dotfill\text{(iii)}.$$

Now the relation between the curves C and C_1 is clearly a reciprocal one: The point $\mathbf{r}$ is at a distance $-a$ along the normal at
$\mathbf{r}_1$, and $\mathbf{t}$ is inclined at an angle $-\alpha$ to $\mathbf{t}_1$. Hence, corresponding to
(iii), we have

$$\left.\begin{aligned}
\cos\alpha &= (1 + a\kappa_1)\frac{ds_1}{ds}\\
\sin\alpha &= -\,a\tau_1\frac{ds_1}{ds}
\end{aligned}\right\} \quad\dotfill\text{(iv)}.$$

On multiplying together corresponding formulae of (iii) and (iv)
we obtain the relations

$$\left.\begin{aligned}
\tau\tau_1 &= \frac{1}{a^2}\sin^2\alpha\\
(1 - a\kappa)(1 + a\kappa_1) &= \cos^2\alpha
\end{aligned}\right\} \quad\dotfill\text{(v)}.$$

The first of these shows that *the torsions of the two curves have
the same sign, and their product is constant.* This theorem is due to

Schell. The result contained in the second formula may be expressed as follows: *If P, P_1 are corresponding points on two conjugate Bertrand curves, and O, O_1 their centres of curvature, the cross ratio of the range (POP_1O_1) is constant and equal to $\sec^2 a$.* This theorem is due to Mannheim.

Ex. 1. By differentiating the equation

$$(1 - a\kappa) \sin a + a\tau \cos a = 0,$$

deduce the following results:

For a curve of constant curvature the conjugate is the locus of its centre of curvature.

A curve of constant torsion coincides with its conjugate.

Ex. 2. Show that a plane curve admits an infinity of conjugates, all *parallel* to the given curve.

Prove also that the only other curve which has more than one conjugate is the circular helix, the conjugates being also circular helices on coaxial cylinders.

EXAMPLES II

1. The principal normal to a curve is normal to the locus of the centre of curvature at points for which the value of κ is stationary.

2. The normal plane to the locus of the centre of circular curvature of a curve C bisects the radius of spherical curvature at the corresponding point of C.

3. The binormal at a point P of a given curve is the limiting position of the common perpendicular to the tangents at P and a consecutive point of the curve.

4. For a curve drawn on a sphere the centre of curvature at any point is the foot of the perpendicular from the centre of the sphere upon the osculating plane at the point.

5. Prove that, in order that the principal normals of a curve be binormals of another, the relation

$$a(\kappa^2 + \tau^2) = \kappa$$

must hold, where a is constant.

6. If there is a one-to-one correspondence between the points of two curves, and the tangents at corresponding points are parallel, show that the principal normals are parallel, and therefore also the binormals. Prove also that

$$\frac{\kappa_1}{\kappa} = \frac{ds}{ds_1} = \frac{\tau_1}{\tau}.$$

Two curves so related are said to be deducible from each other by a *Combescure transformation*.

7. A curve is traced on a right circular cone so as to cut all the generating lines at a constant angle. Show that its projection on the plane of the base is an equiangular spiral.

8. Find the curvature and torsion of the curve in the preceding example.

9. A curve is drawn on a right circular cone, everywhere inclined at the same angle a to the axis. Prove that $\kappa = \tau \tan a$.

10. Determine the curves which have a given curve C as the locus of the centre of spherical curvature.

If C_1 is a curve with this property then, by Art. 5, $\mathbf{r}_1$ lies in the osculating plane of C at $\mathbf{r}$. Thus

$$\mathbf{r}_1 = \mathbf{r} + l\mathbf{t} + m\mathbf{n}.$$

Further, the tangent to C_1 is parallel to $\mathbf{b}$. Hence show that

$$l' = \kappa m - 1,$$
$$m' = -\kappa l.$$

Integrate the equations, and show that there is a double infinitude of curves with the required property.

11. On the binormal of a curve of *constant torsion* τ a point Q is taken at a constant distance c from the curve. Show that the binormal to the locus of Q is inclined to the binormal of the given curve at an angle

$$\tan^{-1} \frac{c\tau^2}{\kappa \sqrt{c^2\tau^2 + 1}}.$$

12. On the *tangent* to a given curve a point Q is taken at a constant distance c from the point of contact. Prove that the curvature κ_1 of the locus of Q is given by

$$\kappa_1^2 (1 + c^2\kappa^2)^3 = c^2\kappa^2\tau^2 (1 + c^2\kappa^2) + (\kappa + c\kappa' + c^2\kappa^3)^2.$$

13. On the *binormal* to a given curve a point Q is taken at a constant distance c from the curve. Prove that the curvature κ_1 of the locus of Q is given by

$$\kappa_1^2 (1 + c^2\tau^2)^3 = c^2\tau^4 (1 + c^2\tau^2) + (\kappa - c\tau' + c^2\kappa\tau^2)^2.$$

14. Prove that the curvature κ_1 of the locus of the centre of (circular) curvature of a given curve is given by

$$\kappa_1^2 = \left\{ \frac{\rho^2\sigma}{R^3} \frac{d}{ds}\left(\frac{\sigma\rho'}{\rho}\right) - \frac{1}{R} \right\}^2 + \frac{\rho'^2\sigma^4}{\rho^2 R^4},$$

where the symbols have their usual meanings.

CHAPTER II

ENVELOPES. DEVELOPABLE SURFACES

13. Surfaces. We have seen that a curve is the locus of a point whose coordinates x, y, z are functions of a single parameter. We now define a surface as the locus of a point whose coordinates are functions of *two independent parameters* u, v. Thus

$$x = f_1(u, v), \quad y = f_2(u, v), \quad z = f_3(u, v) \quad \ldots\ldots\ldots(1)$$

are *parametric equations of a surface*. In particular cases one, or even two, of the functions may involve only a single parameter. If now u, v are eliminated from the equations (1) we obtain a relation between the coordinates which may be written

$$F(x, y, z) = 0 \quad \ldots\ldots\ldots\ldots\ldots\ldots(2).$$

This is the oldest form of the equation of a surface. The two-parametric representation of a surface as given in (1) is due to Gauss. In subsequent chapters it will form the basis of our investigation. But for the discussion in the present chapter the form (2) of the equation of a surface will prove more convenient.

14. Tangent plane. Normal. Consider any curve drawn on the surface

$$F(x, y, z) = 0.$$

Let s be the arc-length measured from a fixed point up to the current point (x, y, z). Then, since the function F has the same value at all points of the surface, it remains constant along the curve as s varies. Thus

$$\frac{\partial F}{\partial x}\frac{dx}{ds} + \frac{\partial F}{\partial y}\frac{dy}{ds} + \frac{\partial F}{\partial z}\frac{dz}{ds} = 0,$$

which we may write more briefly

$$F_x x' + F_y y' + F_z z' = 0.$$

Now the vector (x', y', z') is the unit tangent to the curve at the point (x, y, z); and the last equation shows that it is perpendicular to the vector (F_x, F_y, F_z). The tangent to any curve drawn on a surface is called a *tangent line* to the surface. Thus all tangent lines to the surface at the point (x, y, z) are perpendicular to the vector (F_x, F_y, F_z), and therefore lie in the plane through (x, y, z) perpendicular to this vector. This plane is called the *tangent plane* to the surface at that point, and the normal to the plane at the point of contact is called the *normal* to the surface at that point. Since the line joining any point (X, Y, Z) on the tangent plane to the point of contact is perpendicular to the normal, it follows that

$$(X - x)\frac{\partial F}{\partial x} + (Y - y)\frac{\partial F}{\partial y} + (Z - z)\frac{\partial F}{\partial z} = 0 \quad \ldots\ldots(3).$$

This is the equation of the tangent plane. Similarly if (X, Y, Z) is a current point on the normal, we have

$$\frac{X - x}{\dfrac{\partial F}{\partial x}} = \frac{Y - y}{\dfrac{\partial F}{\partial y}} = \frac{Z - z}{\dfrac{\partial F}{\partial z}} \quad \ldots\ldots\ldots\ldots(4).$$

These are the equations of the normal at the point (x, y, z).

Ex. 1. Prove that the tangent plane to the surface $xyz = a^3$, and the coordinate planes, bound a tetrahedron of constant volume.

Ex. 2. Show that the sum of the squares of the intercepts on the coordinate axes made by the tangent plane to the surface

$$x^{\frac{2}{3}} + y^{\frac{2}{3}} + z^{\frac{2}{3}} = a^{\frac{2}{3}}$$

is constant.

Ex. 3. At points common to the surface

$$a(yz + zx + xy) = xyz$$

and a sphere whose centre is the origin, the tangent plane to the surface makes intercepts on the axes whose sum is constant.

Ex. 4. The normal at a point P of the ellipsoid

$$\frac{x^2}{a^2} + \frac{y^2}{b^2} + \frac{z^2}{c^2} = 1$$

meets the coordinate planes in G_1, G_2, G_3. Prove that the ratios

$$PG_1 : PG_2 : PG_3$$

are constant.

Ex. 5. Any tangent plane to the surface

$$a(x^2 + y^2) + xyz = 0$$

meets it again in a conic whose projection on the plane of xy is a rectangular hyperbola.

15. Envelope. Characteristics. An equation of the form

$$F(x, y, z, a) = 0 \ldots\ldots\ldots\ldots\ldots\ldots\ldots (5),$$

in which a is constant, represents a surface. If the value of the constant is altered, so in general is the surface. The infinitude of surfaces, which correspond to the infinitude of values that may be assigned to a, is called a *family* of surfaces with parameter a. On any one surface the value of a is constant; it changes, however, from one surface to another. This parameter has then a different significance from that of the parameters u, v in Art. 13. These relate to a single surface, and vary from point to point of that surface. They are curvilinear coordinates of a point on a single surface. The parameter a, however, determines a particular member of a family of surfaces, and has the same value at all points of that member.

The curve of intersection of two surfaces of the family corresponding to the parameter values a and $a + \delta a$ is determined by the equations

$$F(x, y, z, a) = 0, \quad F(x, y, z, a + \delta a) = 0,$$

and therefore by the equations

$$F(a) = 0, \quad \frac{F(a + \delta a) - F(a)}{\delta a} = 0,$$

in which, for the sake of brevity, we have written $F(a)$ instead of $F(x, y, z, a)$, and so on. If now we make δa tend to zero, the curve becomes the curve of intersection of consecutive members of the family, and its defining equations become

$$F(a) = 0, \quad \frac{\partial}{\partial a} F(a) = 0 \ \ldots\ldots\ldots\ldots (6).$$

This curve is called the *characteristic* of the surface for the parameter value a. As the parameter varies we obtain a family of such characteristics, and their locus is called the *envelope* of the family of surfaces. It is the surface whose equation is obtained by eliminating a from the two equations (6).

Two surfaces are said to *touch* each other at a common point when they have the same tangent plane, and therefore the same normal, at that point. We shall now prove that *the envelope touches each member of the family of surfaces at all points of its character-istic.*

The characteristic corresponding to the parameter value a lies both on the surface with the same parameter value and on the envelope. Thus all points of the characteristic are common to the surface and the envelope. The normal to the surface

$$F(x, y, z, a) = 0$$

is parallel to the vector

$$\left(\frac{\partial F}{\partial x},\ \frac{\partial F}{\partial y},\ \frac{\partial F}{\partial z}\right) \quad\dots\dots\dots\dots\dots(\text{i}).$$

The equation of the envelope is obtained by eliminating a from the equations (6). The envelope is therefore represented by $F(x, y, z, a) = 0$, provided a is regarded as a function of x, y, z given by

$$\frac{\partial}{\partial a} F(x, y, z, a) = 0.$$

The normal to the envelope is then parallel to the vector

$$\left(\frac{\partial F}{\partial x} + \frac{\partial F}{\partial a}\frac{\partial a}{\partial x},\ \frac{\partial F}{\partial y} + \frac{\partial F}{\partial a}\frac{\partial a}{\partial y},\ \frac{\partial F}{\partial z} + \frac{\partial F}{\partial a}\frac{\partial a}{\partial z}\right),$$

which, in virtue of the preceding equation, is the same as the vector (i). Thus, at all common points, the surface and the envelope have the same normal, and therefore the same tangent plane; so that they touch each other at all points of the characteristic.

Ex. 1. The envelope of the family of paraboloids
$$x^2 + y^2 = 4a(z - a)$$
is the circular cone $\quad x^2 + y^2 = z^2.$

Ex. 2. Spheres of constant radius b have their centres on the fixed circle $x^2 + y^2 = a^2$, $z = 0$. Prove that their envelope is the surface
$$(x^2 + y^2 + z^2 + a^2 - b^2)^2 = 4a^2(x^2 + y^2).$$

Ex. 3. The envelope of the family of surfaces
$$F(x, y, z, a, b) = 0,$$
in which the parameters a, b are connected by the equation
$$f(a, b) = 0,$$
is found by eliminating a and b from the equations
$$F = 0, \quad f = 0, \quad \frac{F_a}{f_a} = \frac{F_b}{f_b}.$$

16. Edge of regression. The locus of the ultimate intersections of consecutive characteristics of a one-parameter family of surfaces is called the *edge of regression*. It is easy to show that *each characteristic touches the edge of regression*, that is to say, the two curves have the same tangent at their common point. For if A, B, C are three consecutive characteristics, A and B intersecting at P, and B and C at Q, these two points are consecutive points on the characteristic B and also on the edge of regression. Hence ultimately, as A and C tend to coincidence with B, the chord PQ becomes a common tangent to the characteristic and to the edge of regression.

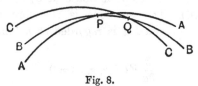

Fig. 8.

The same may be proved analytically as follows. The characteristic with parameter value a is the curve of intersection of the surfaces

$$F(a) = 0, \quad \frac{\partial}{\partial a} F(a) = 0 \quad \dots\dots\dots\dots(6).$$

The tangent to the characteristic at any point is therefore perpendicular to the normals to both surfaces at this point. It is therefore perpendicular to each of the vectors

$$\left(\frac{\partial F}{\partial x}, \frac{\partial F}{\partial y}, \frac{\partial F}{\partial z}\right) \text{ and } \left(\frac{\partial^2 F}{\partial x \partial a}, \frac{\partial^2 F}{\partial y \partial a}, \frac{\partial^2 F}{\partial z \partial a}\right) \quad \dots\dots(7).$$

The equations of the consecutive characteristic, with the parameter value $a + da$, are

$$F + \frac{\partial F}{\partial a} da = 0, \quad \frac{\partial F}{\partial a} + \frac{\partial^2 F}{\partial a^2} da = 0.$$

Hence, for its point of intersection with (6), since all four equations must be satisfied, we have

$$F(a) = 0, \quad \frac{\partial}{\partial a} F(a) = 0, \quad \frac{\partial^2}{\partial a^2} F(a) = 0 \quad \dots\dots\dots(8).$$

The equations of the edge of regression are obtained by eliminating a from these three equations. We may then regard the edge of

regression as the curve of intersection of the surfaces (6), in which
a is now a function of x, y, z given by

$$\frac{\partial^2}{\partial a^2} F(a) = 0.$$

Thus the tangent to the edge of regression, being perpendicular
to the normals to both surfaces, is perpendicular to each of the
vectors

$$\left(\frac{\partial F}{\partial x} + \frac{\partial F}{\partial a} \frac{\partial a}{\partial x}, \quad \frac{\partial F}{\partial y} + \frac{\partial F}{\partial a} \frac{\partial a}{\partial y}, \quad \frac{\partial F}{\partial z} + \frac{\partial F}{\partial a} \frac{\partial a}{\partial z} \right)$$

and

$$\left(\frac{\partial^2 F}{\partial x \partial a} + \frac{\partial^2 F}{\partial a^2} \frac{\partial a}{\partial x}, \quad \frac{\partial^2 F}{\partial y \partial a} + \frac{\partial^2 F}{\partial a^2} \frac{\partial a}{\partial y}, \quad \ldots \ldots \right),$$

which, in virtue of the equations (8), are the same as the vectors
(7). Thus the tangent to the edge of regression is parallel to the
tangent to the characteristic, and the two curves therefore touch
at their common point.

Ex. 1. Find the envelope of the family of planes
$$3a^2 x - 3ay + z = a^3,$$
and show that its edge of regression is the curve of intersection of the surfaces
$xz = y^2$, $xy = z$.

Ex. 2. Find the edge of regression of the envelope of the family of planes
$$x \sin \theta - y \cos \theta + z = a\theta,$$
θ being the parameter.

Ex. 3. Find the envelope of the family of cones
$$(ax + x + y + z - 1)(ay + z) = ax(x + y + z - 1),$$
a being the parameter.

Ex. 4. Prove that the characteristics of the family of osculating spheres
of a twisted curve are its circles of curvature, and the edge of regression is the
curve itself.

Ex. 5. Find the envelope and the edge of regression of the spheres which
pass through a fixed point and whose centres lie on a given curve.

Ex. 6. Find the envelope and the edge of regression of the family of
ellipsoids
$$c^2 \left(\frac{x^2}{a^2} + \frac{y^2}{b^2} \right) + \frac{z^2}{c^2} = 1,$$
where c is the parameter.

17. Developable surfaces. An important example of the
preceding theory is furnished by a one-parameter family of planes.
In this case the characteristics, being the intersections of consecu-

tive planes, are straight lines. These straight lines are called the *generators* of the envelope, and the envelope is called a *developable surface,* or briefly a *developable.* The reason for the name lies in the fact that the surface may be unrolled or developed into a plane without stretching or tearing. For, since consecutive generators are coplanar, the plane containing the first and second of the family of generators may be turned about the second till it coincides with the plane containing the second and third ; then this common plane may be turned about the third till it coincides with the plane containing the third and the fourth; and so on. In this way the whole surface may be developed into a plane.

Since each plane of the family touches the envelope along its characteristic, it follows that *the tangent plane to a developable surface is the same at all points of a generator.* The edge of regression of the developable is the locus of the intersections of consecutive generators, and is touched by each of the generators. Moreover, since consecutive generators are consecutive tangents to the edge of regression, the osculating plane of this curve is that plane of the family which contains these generators. But this plane touches the developable. Hence *the osculating plane of the edge of regression at any point is the tangent plane to the developable at that point.*

Suppose the equation of a surface is given in Monge's form,

$$z = f(x, y) \quad \dots\dots\dots\dots\dots\dots\dots(9),$$

and we require the condition that the surface may be a developable. The equation of the tangent plane at the point (x, y, z) is

$$Z - z = (X - x)\frac{\partial f}{\partial x} + (Y - y)\frac{\partial f}{\partial y},$$

and, in order that this may be expressible in terms of a single parameter, there must be some relation between f_x and f_y, which we may write

$$\frac{\partial f}{\partial x} = \phi\left(\frac{\partial f}{\partial y}\right).$$

On differentiation this gives

$$\frac{\partial^2 f}{\partial x^2} = \phi'\left(\frac{\partial f}{\partial y}\right)\frac{\partial^2 f}{\partial x \partial y},$$

$$\frac{\partial^2 f}{\partial x \partial y} = \phi'\left(\frac{\partial f}{\partial y}\right)\frac{\partial^2 f}{\partial y^2},$$

and from these it follows that

$$\frac{\partial^2 f}{\partial x^2}\frac{\partial^2 f}{\partial y^2} = \left(\frac{\partial^2 f}{\partial x \partial y}\right)^2 \quad \dots\dots\dots\dots(10).$$

This is the required condition that (9) may represent a developable surface.

Ex. Prove that the surface $xy = (z-c)^2$ is a developable.

DEVELOPABLES ASSOCIATED WITH A CURVE

18. Osculating developable. The principal planes of a twisted curve at a current point P are the osculating plane, which is parallel to **t** and **n**, the normal plane, which is parallel to **n** and **b**, and the rectifying plane, which is parallel to **b** and **t**. The equations of these planes contain only a single parameter, which is usually the arc-length s; and the envelopes of the planes are therefore developable surfaces.

The envelope of the osculating plane is called the *osculating developable*. Since the intersections of consecutive osculating planes are the tangents to the curve, it follows that the tangents are the generators of the developable. And consecutive tangents intersect at a point on the curve; so that the curve itself is the edge of regression of the osculating developable.

The same may be proved analytically as follows. At a point **r** on the curve the equation of the osculating plane is

$$(\mathbf{R} - \mathbf{r})\cdot\mathbf{b} = 0 \quad \dots\dots\dots\dots(11),$$

where **r** and **b** are functions of s. On differentiating with respect to s we have $\quad -\mathbf{t}\cdot\mathbf{b} - \tau(\mathbf{R} - \mathbf{r})\cdot\mathbf{n} = 0,$

that is $\quad\quad\quad (\mathbf{R} - \mathbf{r})\cdot\mathbf{n} = 0 \quad \dots\dots\dots\dots(12),$

which is the equation of the rectifying plane. Thus the characteristic, being given by (11) and (12), is the intersection of the osculating and rectifying planes, and is therefore the tangent to the curve at **r**. To find the edge of regression we differentiate (12) and obtain $\quad (\mathbf{R} - \mathbf{r})\cdot(\tau\mathbf{b} - \kappa\mathbf{t}) = 0 \quad \dots\dots\dots(13).$

For a point on the edge of regression all three equations (11), (12) and (13) are satisfied. Hence $(\mathbf{R} - \mathbf{r})$ vanishes identically, and the curve itself is the edge of regression.

Ex. Find the osculating developable of the circular helix.

19. Polar developable. The envelope of the normal plane of a twisted curve is called the *polar developable*, and its generators are called the *polar lines*. Thus the polar line for the point P is the intersection of consecutive normal planes at P. The equation of the normal plane is

$$(\mathbf{R} - \mathbf{r})\cdot\mathbf{t} = 0 \qquad\dots\dots\dots\dots(14),$$

where $\mathbf{r}$ and $\mathbf{t}$ are functions of s. Differentiating with respect to s we find

$$\kappa(\mathbf{R} - \mathbf{r})\cdot\mathbf{n} - \mathbf{t}\cdot\mathbf{t} = 0,$$

which may be written $(\mathbf{R} - \mathbf{r} - \rho\mathbf{n})\cdot\mathbf{n} = 0$ $\dots\dots\dots(15)$. This equation represents a plane through the centre of curvature perpendicular to the principal normal. It intersects the normal plane in a straight line through the centre of curvature parallel to the binormal (Fig. 4). Thus *the polar line is the axis of the circle of curvature*. On differentiating (15) we obtain the third equation for the edge of regression,

$$(\mathbf{R} - \mathbf{r})\cdot(\tau\mathbf{b} - \kappa\mathbf{t}) = \rho',$$

which, in virtue of (14), may be written

$$(\mathbf{R} - \mathbf{r})\cdot\mathbf{b} = \sigma\rho' \qquad\dots\dots\dots\dots(16).$$

From the three equations (14), (15) and (16) it follows that

$$\mathbf{R} - \mathbf{r} = \rho\mathbf{n} + \sigma\rho'\mathbf{b},$$

so that the point $\mathbf{R}$ coincides with the centre of spherical curvature. Thus *the edge of regression of the polar developable is the locus of the centre of spherical curvature*. The tangents to this locus are the polar lines, which are the generators of the developable.

20. Rectifying developable. The envelope of the rectifying plane of a curve is called the *rectifying developable*, and its generators are the *rectifying lines*. Thus the rectifying line at a point P of the curve is the intersection of consecutive rectifying planes. The equation of the rectifying plane at the point $\mathbf{r}$ is

$$(\mathbf{R} - \mathbf{r})\cdot\mathbf{n} = 0 \qquad\dots\dots\dots\dots(17),$$

where $\mathbf{r}$ and $\mathbf{n}$ are functions of s. The other equation of the rectifying line is got by differentiating with respect to s, thus obtaining

$$(\mathbf{R} - \mathbf{r})\cdot(\tau\mathbf{b} - \kappa\mathbf{t}) = 0 \qquad\dots\dots\dots\dots(18).$$

From these equations it follows that the rectifying line passes through the point $\mathbf{r}$ on the curve, and is perpendicular to both $\mathbf{n}$

and $(\tau\mathbf{b} - \kappa\mathbf{t})$. Hence it is parallel to the vector $(\tau\mathbf{t} + \kappa\mathbf{b})$, and is therefore inclined to the tangent at an angle ϕ such that

$$\tan \phi = \frac{\kappa}{\tau}. \dots\dots\dots\dots\dots\dots(19).$$

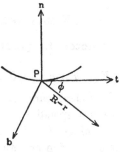

Fig. 9.

To find the edge of regression we differentiate (18) and, in virtue of (17), we obtain

$$(\mathbf{R} - \mathbf{r})\cdot(\tau'\mathbf{b} - \kappa'\mathbf{t}) + \kappa = 0 \dots\dots\dots\dots(20).$$

Further, since the rectifying line is parallel to $\tau\mathbf{t} + \kappa\mathbf{b}$, the point $\mathbf{R}$ on the edge of regression is such that

$$(\mathbf{R} - \mathbf{r}) = l\,(\tau\mathbf{t} + \kappa\mathbf{b}),$$

where l is some number. On substitution of this in (20) we find

$$l = \frac{\kappa}{\kappa'\tau - \kappa\tau'}.$$

Thus the point on the edge of regression corresponding to the point $\mathbf{r}$ on the curve is

$$\mathbf{R} = \mathbf{r} + \frac{\kappa\,(\tau\mathbf{t} + \kappa\mathbf{b})}{\kappa'\tau - \kappa\tau'} \dots\dots\dots\dots\dots(21).$$

The reason for the term "rectifying" applied to this developable lies in the fact that, when the surface is developed into a plane by unfolding about consecutive generators, the original curve becomes a straight line. The truth of this statement will appear later when we consider the properties of "geodesics" on a surface.

We may notice in passing that, if the given curve is a helix, κ/τ is constant, and the angle ϕ of (19) is equal to the angle β of Art. 8. Thus the rectifying lines are the generators of the cylinder on which the helix is drawn, and the rectifying developable is the cylinder itself.

Ex. Prove that the rectifying developable of a curve is the polar developable of its involutes, and conversely.

Two-parameter Family of Surfaces

21. Envelope. Characteristic points. An equation of the form

$$F(x, y, z, a, b) = 0 \ldots\ldots\ldots\ldots\ldots\ldots(22),$$

in which a and b are independent parameters, represents a doubly infinite family of surfaces, corresponding to the infinitude of values of a and the infinitude of values of b. On any one surface both a and b are constant. The curve of intersection of the surface whose parameter values are a, b with the consecutive surface whose parameter values are $a + da$, $b + db$ is given by the equations

$$F(a, b) = 0, \quad F(a + da, \ b + db) = 0,$$

or by the equations

$$F(a, b) = 0, \quad F(a, b) + \frac{\partial}{\partial a} F(a, b)\, da + \frac{\partial}{\partial b} F(a, b)\, db = 0.$$

This curve depends on the ratio $da : db$; but for all values of this ratio it passes through the point or points given by

$$F(a, b) = 0, \quad \frac{\partial}{\partial a} F(a, b) = 0, \quad \frac{\partial}{\partial b} F(a, b) = 0 \ \ldots(23).$$

These are called *characteristic points*, and the locus of the characteristic points is called the *envelope* of the family of surfaces. The equation of the envelope is obtained by eliminating a and b from the equations (23).

Each characteristic point is common to the envelope and one surface of the family; and we can prove that *the envelope touches each surface at the characteristic point* (*or points*). The normal to a surface of the family at the point (x, y, z) is parallel to the vector

$$\left(\frac{\partial F}{\partial x}, \ \frac{\partial F}{\partial y}, \ \frac{\partial F}{\partial z}\right) \ \ldots\ldots\ldots\ldots\ldots\ldots(24).$$

The equation of the envelope is got by eliminating a and b from the equations (23). We may therefore take $F(x, y, z, a, b) = 0$ as the equation of the envelope, provided we regard a, b as functions of x, y, z given by

$$\frac{\partial F}{\partial a} = 0, \quad \frac{\partial F}{\partial b} = 0 \quad \ldots\ldots\ldots\ldots\ldots(25).$$

Then the normal to the envelope is parallel to the vector

$$\left(\frac{\partial F}{\partial x} + \frac{\partial F}{\partial a}\frac{\partial a}{\partial x} + \frac{\partial F}{\partial b}\frac{\partial b}{\partial x}, \quad \frac{\partial F}{\partial y} + \frac{\partial F}{\partial a}\frac{\partial a}{\partial y} + \frac{\partial F}{\partial b}\frac{\partial b}{\partial y}, \quad \ldots \right),$$

which, in virtue of (25), is the same as the vector (24). Thus the envelope has the same normal, and therefore the same tangent plane, as a surface of the family at the characteristic point. The contact property is thus established.

Ex. 1. Show that the envelope of the plane

$$\frac{x}{a}\cos\theta\sin\phi + \frac{y}{b}\sin\theta\sin\phi + \frac{z}{c}\cos\phi = 1,$$

where θ, ϕ are independent parameters, is the ellipsoid

$$\frac{x^2}{a^2} + \frac{y^2}{b^2} + \frac{z^2}{c^2} = 1.$$

Ex. 2. Prove that the envelope of a plane which forms with the coordinate planes a tetrahedron of constant volume is a surface $xyz = \text{const.}$

Ex. 3. The envelope of a plane, the sum of the squares of whose intercepts on the axes is constant, is a surface

$$x^{\frac{3}{2}} + y^{\frac{3}{2}} + z^{\frac{3}{2}} = \text{const.}$$

Ex. 4. The envelope of the plane

$$(u-v)\,bcx + (1+uv)\,cay + (1-uv)\,abz = abc\,(u+v),$$

where u, v are parameters, is the hyperboloid

$$\frac{x^2}{a^2} + \frac{y^2}{b^2} - \frac{z^2}{c^2} = 1.$$

Ex. 5. Prove that the envelope of the surface $F(x, y, z, a, b, c) = 0$, where a, b, c are parameters connected by the relation $f(a, b, c) = 0$, is obtained by eliminating a, b, c from the equations

$$F = 0, \quad f = 0, \quad \frac{F_a}{f_a} = \frac{F_b}{f_b} = \frac{F_c}{f_c}.$$

Ex. 6. The envelope of the plane $lx + my + nz = p$, where

$$p^2 = a^2 l^2 + b^2 m^2 + c^2 n^2,$$

is an ellipsoid.

EXAMPLES III

1. Find the envelope of the planes through the centre of an ellipsoid and cutting it in sections of constant area.

2. Through a fixed point on a given circle chords are drawn. Find the envelope of the spheres on these chords as diameters.

3. A plane makes intercepts a, b, c on the coordinate axes such that

$$\frac{1}{a^2}+\frac{1}{b^2}+\frac{1}{c^2}=\frac{1}{k^2}.$$

Prove that its envelope is a conicoid with equi-conjugate diameters along the axes.

4. A fixed point O on the x-axis is joined to a variable point P on the yz-plane. Find the envelope of the plane through P at right angles to OP.

5. Find the envelope of the plane

$$\frac{x}{a+u}+\frac{y}{b+u}+\frac{z}{c+u}=1,$$

where u is the parameter, and determine the edge of regression.

6. The envelope of a plane, such that the sum of the squares of its distances from n given points is constant, is a conicoid with centre at the centroid of the given points.

7. A fixed point O is joined to a variable point P on a given spherical surface. Find the envelope of the plane through P at right angles to OP.

8. A sphere of constant radius a moves with its centre on a given twisted curve. Prove that the characteristic for any position of the sphere is its great circle by the normal plane to the curve. Show also that, if the radius of curvature ρ of the curve is less than a, the edge of regression consists of two branches, on which the current point is

$$\mathbf{r}+\rho\mathbf{n}\pm\sqrt{a^2-\rho^2}\,\mathbf{b}$$

The envelope is called a *canal surface*.

9. Show that the radius of curvature of the edge of regression of the rectifying developable (Art. 20) is equal to $\operatorname{cosec}\phi\,\dfrac{d}{d\phi}\left(\sin^2\phi\,\dfrac{ds}{d\phi}\right)$, where $\tan\phi=\dfrac{\sigma}{\rho}$, and that the radius of torsion is equal to

$$-\rho\,\frac{d}{ds}\left(\sin^2\phi\,\frac{ds}{d\phi}\right).$$

CHAPTER III

CURVILINEAR COORDINATES ON A SURFACE. FUNDAMENTAL MAGNITUDES

22. Curvilinear coordinates. We have seen that a surface may be regarded as the locus of a point whose position vector $\mathbf{r}$ is a function of two independent parameters u, v. The Cartesian coordinates x, y, z of the point are then known functions of u, v; and the elimination of the two parameters leads to a single relation between x, y, z which is usually called the *equation of the surface*. We shall confine our attention to surfaces, or portions of surfaces, which present no singularities of any kind.

Any relation between the parameters, say $f(u, v) = 0$, represents a curve on the surface. For $\mathbf{r}$ then becomes a function of only one independent parameter, so that the locus of the point is a curve. In particular the curves on the surface, along which one of the

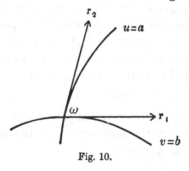

Fig. 10.

parameters remains constant, are called the *parametric curves*. The surface can be mapped out by a doubly infinite set of parametric curves, corresponding to the infinitude of values that can be assigned to each of the parameters. The parameters u, v thus constitute a system of *curvilinear coordinates* for points on the surface, the position of the point being determined by the values of u and v.

4—2

Suppose, for example, that we are dealing with the surface of a sphere of radius a, and that three mutually perpendicular diameters are chosen as coordinate axes. The latitude λ of a point P on the surface may be defined as the inclination of the radius through P to the xy plane, and the longitude ϕ as the inclination of the plane containing P and the z-axis to the zx plane. Then the coordinates of P are given by

$$x = a \cos \lambda \cos \phi, \quad y = a \cos \lambda \sin \phi, \quad z = a \sin \lambda.$$

Thus λ and ϕ may be taken as parameters for the surface. The parametric curves $\lambda = $ const. are the small circles called parallels of latitude; the curves $\phi = $ const. are the great circles called meridians of longitude. As these two systems of curves cut each other at right angles, we say the parametric curves are *orthogonal*.

As another example consider the osculating developable of a twisted curve. The generators of this surface are the tangents to the curve. Hence the position vector of a point on the surface is given by

$$\mathbf{R} = \mathbf{r} + u\mathbf{t},$$

where u is the distance of the point from the curve measured along the tangent at the point $\mathbf{r}$. But $\mathbf{r}$, $\mathbf{t}$ are functions of the arc-length s of the given curve. Hence s, u may be taken as parameters for the osculating developable. The parametric curves $s = $ const. are the generators; and the curves $u = $ const. cut the tangents at a constant distance from the given curve.

If the equation of the surface is given in Monge's form

$$z = f(x, y),$$

the coordinates x, y may be taken as parameters. In this case the parametric curves are the intersections of the surface with the planes $x = $ const. and $y = $ const.

Ex. 1. On the surface of revolution

$$x = u \cos \phi, \quad y = u \sin \phi, \quad z = f(u),$$

what are the parametric curves $u = $ const., and what are the curves $\phi = $ const.?

Ex. 2. On the right helicoid given by

$$x = u \cos \phi, \quad y = u \sin \phi, \quad z = c\phi,$$

show that the parametric curves are circular helices and straight lines.

Ex. 3. On the hyperboloid of one sheet

$$\frac{x}{a} = \frac{\lambda + \mu}{1 + \lambda\mu}, \quad \frac{y}{b} = \frac{1 - \lambda\mu}{1 + \lambda\mu}, \quad \frac{z}{c} = \frac{\lambda - \mu}{1 + \lambda\mu},$$

the parametric curves are the generators. What curves are represented by $\lambda = \mu$, and by $\lambda\mu = \text{const.}$?

23. First order magnitudes. The suffix 1 will be used to indicate partial differentiation with respect to u, and the suffix 2 partial differentiation with respect to v. Thus

$$\mathbf{r}_1 = \frac{\partial \mathbf{r}}{\partial u}, \quad \mathbf{r}_2 = \frac{\partial \mathbf{r}}{\partial v},$$

$$\mathbf{r}_{11} = \frac{\partial^2 \mathbf{r}}{\partial u^2}, \quad \mathbf{r}_{12} = \frac{\partial^2 \mathbf{r}}{\partial u \partial v}, \quad \mathbf{r}_{22} = \frac{\partial^2 \mathbf{r}}{\partial v^2},$$

and so on. The vector $\mathbf{r}_1$ is tangential to the curve $v = \text{const.}$ at the point $\mathbf{r}$; for its direction is that of the displacement $d\mathbf{r}$ due to a variation du in the first parameter only. We take the positive direction along the parametric curve $v = \text{const.}$ as that for which u increases. This is the direction of the vector $\mathbf{r}_1$ (Fig. 10). Similarly $\mathbf{r}_2$ is tangential to the curve $u = \text{const.}$ in the positive sense, which corresponds to increase of v.

Consider two neighbouring points on the surface, with position vectors $\mathbf{r}$ and $\mathbf{r} + d\mathbf{r}$, corresponding to the parameter values u, v and $u + du$, $v + dv$ respectively. Then

$$d\mathbf{r} = \frac{\partial \mathbf{r}}{\partial u} du + \frac{\partial \mathbf{r}}{\partial v} dv$$

$$= \mathbf{r}_1 du + \mathbf{r}_2 dv.$$

Since the two points are adjacent points on a curve passing through them, the length ds of the element of arc joining them is equal to their actual distance $|d\mathbf{r}|$ apart. Thus

$$ds^2 = d\mathbf{r}^2 = (\mathbf{r}_1 du + \mathbf{r}_2 dv)^2$$

$$= \mathbf{r}_1^2 du^2 + 2\mathbf{r}_1 \cdot \mathbf{r}_2 du\, dv + \mathbf{r}_2^2 dv^2.$$

If then we write $E = \mathbf{r}_1^2, \quad F = \mathbf{r}_1 \cdot \mathbf{r}_2, \quad G = \mathbf{r}_2^2$ ....(1), we have the formula

$$ds^2 = E du^2 + 2F du\, dv + G dv^2 \quad ....(2).$$

The quantities denoted by E, F, G are called the *fundamental magnitudes of the first order.* They are of the greatest importance, and will occur throughout the remainder of this book. The quantity $EG - F^2$ is positive on a real surface when u and v are real. For $\sqrt{E}$ and $\sqrt{G}$ are the modules of $\mathbf{r}_1$ and $\mathbf{r}_2$; and, if ω denote the angle between these vectors, $F = \sqrt{EG} \cos \omega$, and therefore $EG - F^2$ is positive. We shall use the notation

$$H^2 = EG - F^2 \quad\ldots\ldots\ldots\ldots\ldots\ldots(3),$$

and let H denote the positive square root of this quantity.

The length of an element of the parametric curve $v = $ const. is found from (2) by putting $dv = 0$. Its value is therefore $\sqrt{E}\,du$. The unit vector tangential to the curve $v = $ const. is thus

$$\mathbf{a} = \frac{1}{\sqrt{E}} \frac{\partial \mathbf{r}}{\partial u} = E^{-\frac{1}{2}} \mathbf{r}_1.$$

Similarly the length of an element of the curve $u = $ const. is $\sqrt{G}\,dv$, and the unit tangent to this curve is

$$\mathbf{b} = \frac{1}{\sqrt{G}} \frac{\partial \mathbf{r}}{\partial v} = G^{-\frac{1}{2}} \mathbf{r}_2.$$

The two parametric curves through any point of the surface cut at an angle ω such that

$$\cos \omega = \mathbf{a} \cdot \mathbf{b} = \frac{\mathbf{r}_1 \cdot \mathbf{r}_2}{\sqrt{EG}} = \frac{F}{\sqrt{EG}}$$

Therefore* $$\sin \omega = \sqrt{\frac{EG - F^2}{EG}} = \frac{H}{\sqrt{EG}}$$ $\ldots\ldots\ldots\ldots(4).$

and $$\tan \omega = \frac{H}{F}$$

Also since $$\sin \omega = |\mathbf{a} \times \mathbf{b}| = \frac{1}{\sqrt{EG}} |\mathbf{r}_1 \times \mathbf{r}_2|,$$

it follows that $$|\mathbf{r}_1 \times \mathbf{r}_2| = H \quad\ldots\ldots\ldots\ldots\ldots\ldots(5).$$

The parametric curves will cut at right angles at any point if $F = 0$ at that point; and they will do so at all points if $F = 0$ over the surface. In this case they are said to be *orthogonal.* Thus $F = 0$ *is the necessary and sufficient condition that the parametric curves may form an orthogonal system.*

* See also Note I, p. 263.

Ex. 1. For a surface of revolution (cf. Ex. 1, Art. 22)

$$\mathbf{r} = (u \cos v,\ u \sin v,\ f(u)),$$
$$\mathbf{r}_1 = (\cos v,\ \sin v,\ f'(u)),$$
$$\mathbf{r}_2 = (-u \sin v,\ u \cos v,\ 0);$$
$$\therefore\ E = \mathbf{r}_1{}^2 = 1 + f'^2,$$
$$F = \mathbf{r}_1 \cdot \mathbf{r}_2 = 0,$$
$$G = \mathbf{r}_2{}^2 = u^2.$$

Thus the parametric curves are orthogonal, and

$$ds^2 = (1 + f'^2)\,du^2 + u^2 dv^2.$$

Ex. 2. Calculate the same quantities for the surface in Ex. 2 of the preceding Art.

24. Directions on a surface. Any direction on the surface from a given point (u, v) is determined by the increments du, dv in the parameters for a small displacement in that direction. Let ds be the length of the displacement $d\mathbf{r}$ corresponding to the increments du, dv; and let δs be the length of another displacement $\delta\mathbf{r}$ due to increments δu, δv. Then

$$d\mathbf{r} = \mathbf{r}_1 du + \mathbf{r}_2 dv,$$

and
$$\delta\mathbf{r} = \mathbf{r}_1 \delta u + \mathbf{r}_2 \delta v.$$

The inclination ψ of these directions is then given by

$$ds\,\delta s \cos \psi = d\mathbf{r} \cdot \delta\mathbf{r}$$
$$= E\,du\,\delta u + F\,(du\,\delta v + dv\,\delta u) + G\,dv\,\delta v,$$

and[*]
$$ds\,\delta s \sin \psi = |\,d\mathbf{r} \times \delta\mathbf{r}\,|$$
$$= |\,du\,\delta v - dv\,\delta u\,|\ |\,\mathbf{r}_1 \times \mathbf{r}_2\,|$$
$$= H\,|\,du\,\delta v - dv\,\delta u\,|.$$

The two directions are perpendicular if $\cos \psi = 0$, that is if

$$E\,\frac{du}{dv}\,\frac{\delta u}{\delta v} + F\left(\frac{du}{dv} + \frac{\delta u}{\delta v}\right) + G = 0 \quad\ldots\ldots\ldots\ldots(6).$$

As an important particular case, the angle θ between the direc-

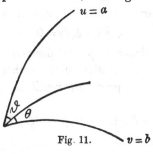

Fig. 11.

$u = a$

$v = b$

tion du, dv and that of the curve $v = $ const. may be deduced from the above results by putting $\delta v = 0$ and $\delta s = \sqrt{E}\,\delta u$. Thus

$$\cos\theta = \frac{1}{\sqrt{E}}\left(E\frac{du}{ds} + F\frac{dv}{ds}\right)$$

and
$$\sin\theta = \frac{H}{\sqrt{E}}\left|\frac{dv}{ds}\right|$$ ....(7).

Similarly its inclination ϑ to the parametric curve $u = $ const. is obtained by putting $\delta u = 0$ and $\delta s = \sqrt{G}\,\delta v$. Thus

$$\cos\vartheta = \frac{1}{\sqrt{G}}\left(F\frac{du}{ds} + G\frac{dv}{ds}\right)$$

and
$$\sin\vartheta = \frac{H}{\sqrt{G}}\left|\frac{du}{ds}\right|$$ ....(8).

The formula (6) leads immediately to the differential equation of the *orthogonal trajectories* of the family of curves given by

$$P\delta u + Q\delta v = 0,$$

where P, Q are functions of u, v. For the given family of curves we have

$$\frac{\delta u}{\delta v} = -\frac{Q}{P},$$

and therefore from (6), if du/dv refers to the orthogonal trajectories, it follows that

$$(EQ - FP)\,du + (FQ - GP)\,dv = 0 \ ....(9).$$

This is the required differential equation. If, instead of the differential equation of the original family of curves, we are given their equation in the form

$$\phi(u, v) = c,$$

where c is an arbitrary constant, it follows that

$$\phi_1\delta u + \phi_2\delta v = 0,$$

the suffixes as usual denoting partial derivatives with respect to u and v. The differential equation of the orthogonal trajectories is then obtained from the preceding result by putting $P = \phi_1$ and $Q = \phi_2$, which gives

$$(E\phi_2 - F\phi_1)\,du + (F\phi_2 - G\phi_1)\,dv = 0 \ ....(10).$$

An equation of the form

$$P\,du^2 + Q\,du\,dv + R\,dv^2 = 0$$

determines two directions on the surface, for it is a quadratic in du/dv. Let the roots of the quadratic be denoted by du/dv and $\delta u/\delta v$. Then

$$\frac{du}{dv} + \frac{\delta u}{\delta v} = -\frac{Q}{P},$$

and

$$\frac{du}{dv}\frac{\delta u}{\delta v} = \frac{R}{P}.$$

On substituting these values in (6) we see that the two directions will be at right angles if

$$ER - FQ + GP = 0 \quad\dots\dots\dots\dots(11).$$

Ex. 1. If ψ is the angle between the two directions given by

$$P\,du^2 + Q\,du\,dv + R\,dv^2 = 0,$$

show that

$$\tan\psi = \frac{H\sqrt{Q^2 - 4PR}}{ER - FQ + GP}.$$

Ex. 2. If the parametric curves are orthogonal, show that the differential equation of lines on the surface cutting the curves $u=$const. at a constant angle β is

$$\frac{du}{dv} = \tan\beta \sqrt{\frac{G}{E}}.$$

Ex. 3. Prove that the differential equations of the curves which bisect the angles between the parametric curves are

$$\sqrt{E}\,du - \sqrt{G}\,dv = 0 \quad\text{and}\quad \sqrt{E}\,du + \sqrt{G}\,dv = 0.$$

25. The normal. The normal to the surface at any point is perpendicular to every tangent line through that point, and is

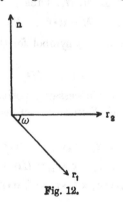

Fig. 12.

therefore perpendicular to each of the vectors r_1 and r_2. Hence it is parallel to the vector $r_1 \times r_2$; and we take the direction of this vector as the positive direction of the normal. The unit vector n parallel to the normal is therefore

$$n = \frac{r_1 \times r_2}{|r_1 \times r_2|} = \frac{r_1 \times r_2}{H} \quad ....(12).$$

This may be called the *unit normal* to the surface. Since it is perpendicular to each of the vectors r_1 and r_2 we have

$$n \cdot r_1 = 0, \quad n \cdot r_2 = 0 \quad ....(13).$$

The scalar triple product of these three vectors has the value

$$[n, r_1, r_2] = n \cdot r_1 \times r_2 = Hn^2 = H \quad ....(14).$$

For the cross products of n with r_1 and r_2 we have

$$r_1 \times n = \frac{1}{H} r_1 \times (r_1 \times r_2) = \frac{1}{H}(Fr_1 - Er_2),$$

and similarly

$$r_2 \times n = \frac{1}{H} r_2 \times (r_1 \times r_2) = \frac{1}{H}(Gr_1 - Fr_2).$$

26. Second order magnitudes. The second derivatives of r with respect to u and v are denoted by

$$r_{11} = \frac{\partial^2 r}{\partial u^2}, \quad r_{12} = \frac{\partial^2 r}{\partial u \partial v}, \quad r_{22} = \frac{\partial^2 r}{\partial v^2}.$$

The *fundamental magnitudes of the second order* are the resolved parts of these vectors in the direction of the normal to the surface. They will be denoted by L, M, N. Thus

$$L = n \cdot r_{11}, \quad M = n \cdot r_{12}, \quad N = n \cdot r_{22}.$$

It will be convenient to have a symbol for the quantity $LN - M^2$. We therefore write

$$T^2 = LN - M^2,$$

though this quantity is not necessarily positive.

We may express L, M, N in terms of scalar triple products of vectors. For

$$[r_1, r_2, r_{11}] = r_1 \times r_2 \cdot r_{11} = Hn \cdot r_{11} = HL.$$

Similarly $[r_1, r_2, r_{12}] = r_1 \times r_2 \cdot r_{12} = Hn \cdot r_{12} = HM,$

and $[r_1, r_2, r_{22}] = r_1 \times r_2 \cdot r_{22} = Hn \cdot r_{22} = HN.$

It will be shown later that the second order magnitudes are intimately connected with the curvature of the surface. We may here observe in passing that they occur in the expression for the length of the perpendicular to the tangent plane from a point on the surface in the neighbourhood of the point of contact. Let $\mathbf{r}$ be the point of contact, P, with parameter values u, v, and $\mathbf{n}$ the unit normal there. The position vector of a neighbouring point Q $(u + du, v + dv)$ on the surface has the value

$$\mathbf{r} + (\mathbf{r}_1 du + \mathbf{r}_2 dv) + \tfrac{1}{2}(\mathbf{r}_{11} du^2 + 2\mathbf{r}_{12} du\, dv + \mathbf{r}_{22} dv^2) + \dots$$

The length of the perpendicular from Q on the tangent plane at P is the projection of the vector PQ on the normal at P, and is therefore equal to

$$\mathbf{n} \cdot (\mathbf{r}_1 du + \mathbf{r}_2 dv) + \tfrac{1}{2}\mathbf{n} \cdot (\mathbf{r}_{11} du^2 + 2\mathbf{r}_{12} du\, dv + \mathbf{r}_{22} dv^2) + \dots$$

In this expression the terms of the first order vanish since $\mathbf{n}$ is at right angles to $\mathbf{r}_1$ and $\mathbf{r}_2$. Hence the length of the perpendicular as far as terms of the second order is

$$\tfrac{1}{2}(L du^2 + 2M du\, dv + N dv^2).$$

Ex. 1. Calculate the fundamental magnitudes for the *right helicoid* given by

$$x = u \cos\phi, \quad y = u \sin\phi, \quad z = c\phi.$$

With u, ϕ as parameters we have

$$\mathbf{r} = (u \cos\phi,\ u \sin\phi,\ c\phi),$$
$$\mathbf{r}_1 = (\cos\phi,\ \sin\phi,\ 0),$$
$$\mathbf{r}_2 = (-u \sin\phi,\ u \cos\phi,\ c).$$

Therefore

$$E = \mathbf{r}_1^2 = 1, \quad F = \mathbf{r}_1 \cdot \mathbf{r}_2 = 0, \quad G = \mathbf{r}_2^2 = u^2 + c^2, \quad H^2 = EG - F^2 = u^2 + c^2.$$

Since $F = 0$ the parametric curves are orthogonal. The unit normal to the surface is

$$\mathbf{n} = \frac{\mathbf{r}_1 \times \mathbf{r}_2}{H} = (c \sin\phi,\ -c \cos\phi,\ u)/H.$$

Further

$$\mathbf{r}_{11} = (0, 0, 0),$$
$$\mathbf{r}_{12} = (-\sin\phi,\ \cos\phi,\ 0),$$
$$\mathbf{r}_{22} = (-u \cos\phi,\ -u \sin\phi,\ 0),$$

so that the second order magnitudes are

$$L = 0, \quad M = -\frac{c}{H}, \quad N = 0.$$

Ex. 2. On the surface given by

$$x = a(u+v), \quad y = b(u-v), \quad z = uv,$$

the parametric curves are straight lines. Further

$$\mathbf{r}_1 = (a, b, v),$$
$$\mathbf{r}_2 = (a, -b, u),$$

and therefore

$$E = a^2 + b^2 + v^2, \quad F = a^2 - b^2 + uv, \quad G = a^2 + b^2 + u^2,$$
$$H^2 = 4a^2 b^2 + a^2 (u-v)^2 + b^2 (u+v)^2.$$

The unit normal is $\mathbf{n} = (bu + bv, \ av - au, \ -2ab)/H.$

Again $\mathbf{r}_{11} = (0, 0, 0),$

$$\mathbf{r}_{12} = (0, 0, 1),$$
$$\mathbf{r}_{22} = (0, 0, 0),$$

and therefore $L = 0, \quad M = -2ab/H, \quad N = 0,$

$$T^2 = LN - M^2 = -4a^2 b^2/H^2.$$

27. Derivatives of n.

Moreover, by means of the fundamental magnitudes we may express the derivatives of $\mathbf{n}$ in terms of $\mathbf{r}_1$ and $\mathbf{r}_2$. Such an expression is clearly possible. For, since $\mathbf{n}$ is a vector of constant length, its first derivatives are perpendicular to $\mathbf{n}$ and therefore tangential to the surface. They are thus parallel to the plane of $\mathbf{r}_1$ and $\mathbf{r}_2$, and may be expressed in terms of these.

We may proceed as follows. Differentiating the relation $\mathbf{n} \cdot \mathbf{r}_1 = 0$ with respect to u we obtain

$$\mathbf{n}_1 \cdot \mathbf{r}_1 + \mathbf{n} \cdot \mathbf{r}_{11} = 0,$$

and therefore $\mathbf{n}_1 \cdot \mathbf{r}_1 = -\mathbf{n} \cdot \mathbf{r}_{11} = -L.$

In the same manner we find

$$\mathbf{n}_1 \cdot \mathbf{r}_2 = -\mathbf{n} \cdot \mathbf{r}_{12} = -M$$
$$\mathbf{n}_2 \cdot \mathbf{r}_1 = -\mathbf{n} \cdot \mathbf{r}_{21} = -M$$
$$\mathbf{n}_2 \cdot \mathbf{r}_2 = -\mathbf{n} \cdot \mathbf{r}_{22} = -N$$

$\ldots\ldots\ldots\ldots\ldots(15).$

Now since $\mathbf{n}_1$ is perpendicular to $\mathbf{n}$ and therefore tangential to the surface we may write

$$\mathbf{n}_1 = a\mathbf{r}_1 + b\mathbf{r}_2,$$

where a and b are to be determined. Forming the scalar products of each side with $\mathbf{r}_1$ and $\mathbf{r}_2$ successively we have

$$-L = aE + bF,$$
$$-M = aF + bG.$$

On solving these equations for a and b, and substituting the values so obtained in the formula for n_1, we find

$$H^2 n_1 = (FM - GL)\, r_1 + (FL - EM)\, r_2 \ldots\ldots\ldots(16).$$

Similarly it may be shown that

$$H^2 n_2 = (FN - GM)\, r_1 + (FM - EN)\, r_2 \ \ldots\ldots(16').$$

If r_2 and r_1 be eliminated in succession from these two equations we obtain expressions for r_1 and r_2 in terms of n_1 and n_2. The reader will easily verify that

$$\left.\begin{aligned} T^2 r_1 &= (FM - EN)\, n_1 + (EM - FL)\, n_2 \\ T^2 r_2 &= (GM - FN)\, n_1 + (FM - GL)\, n_2 \end{aligned}\right\} \ \ldots\ldots(17).$$

These relations could also be proved independently by the same method as that employed in establishing (16).

From the equations (16) and (16') it follows immediately that

$$H^4 n_1 \times n_2 = \{(FM - GL)(FM - EN) - (FL - EM)(FN - GM)\}\, r_1 \times r_2$$
$$= H^3 T^2 n,$$

so that

$$H n_1 \times n_2 = T^2 n \ \ldots\ldots\ldots\ldots\ldots\ldots(18).$$

Thus the scalar triple product

$$[n,\, n_1,\, n_2] = \frac{T^2}{H} n \cdot n = \frac{T^2}{H}.$$

And as a further exercise the reader may easily verify the following relations which will be used later:

$$\left.\begin{aligned} H\,[n,\, n_1,\, r_1] &= EM - FL \\ H\,[n,\, n_1,\, r_2] &= FM - GL \\ H\,[n,\, n_2,\, r_1] &= EN - FM \\ H\,[n,\, n_2,\, r_2] &= FN - GM \end{aligned}\right\} \ \ldots\ldots\ldots\ldots(19).$$

(Cf. Ex. 13 at the end of this chapter.)

28. Curvature of normal section.

It has already been remarked that the quantities L, M, N are connected with the curvature properties of the surface. Consider a normal section of the surface at a given point, that is to say, the section by a plane containing the normal at that point. Such a section is a plane curve whose principal normal is parallel to the normal to the surface. We adopt the convention that *the principal normal to the curve has also the same sense as the unit normal* **n** *to the surface.*

Then the curvature κ_n of the section is positive when the curve is concave on the side toward which $\mathbf{n}$ is directed. Let dashes denote differentiation with respect to the arc-length s of the curve. Then

$$\mathbf{r}'' = \kappa_n \mathbf{n},$$

and therefore $\qquad\qquad \kappa_n = \mathbf{n} \cdot \mathbf{r}''$ ....(20).

But $\qquad\qquad \mathbf{r}' = \mathbf{r}_1 u' + \mathbf{r}_2 v'$

and $\qquad \mathbf{r}'' = \mathbf{r}_1 u'' + \mathbf{r}_2 v'' + \mathbf{r}_{11} u'^2 + 2\mathbf{r}_{12} u'v' + \mathbf{r}_{22} v'^2.$

Consequently, on substituting this value in (20), remembering that $\mathbf{n}$ is perpendicular to $\mathbf{r}_1$ and $\mathbf{r}_2$ we obtain

$$\kappa_n = Lu'^2 + 2Mu'v' + Nv'^2 \quad..............(21).$$

This formula may also be expressed

$$\kappa_n = \frac{L\,du^2 + 2M\,du\,dv + N\,dv^2}{E\,du^2 + 2F\,du\,dv + G\,dv^2}.$$

It gives the curvature of the normal section parallel to the direction du/dv. We may call this briefly the *normal curvature* in that direction. Its reciprocal may be called the *radius of normal curvature*.

Suppose next that the section considered is not a normal section. Then the principal normal to the curve is not parallel to $\mathbf{n}$. It is parallel to $\mathbf{r}''$, and the unit principal normal is $\mathbf{r}''/\kappa$, where κ is the curvature of the section. Let θ be the inclination of the plane of the section to the normal plane which touches the curve at the point considered. Then θ is the angle between $\mathbf{n}$ and the principal normal to the curve. Hence

$$\cos\theta = \mathbf{n} \cdot \mathbf{r}''/\kappa$$

$$= \frac{1}{\kappa}(Lu'^2 + 2Mu'v' + Nv'^2).$$

Now u' has the same value for both sections at the given point, since the two curves touch at that point. Similarly v' is the same for both. Hence the last equation may be written

$$\cos\theta = \kappa_n/\kappa$$

or $\qquad\qquad \kappa_n = \kappa\cos\theta \quad.......................(22).$

This is *Meunier's theorem* connecting the normal curvature in any direction with the curvature of any other section through the same tangent line.

Ex. 1. If L, M, N vanish at all points the surface is a plane.

Ex. 2. A real surface for which the equations

$$\frac{E}{L} = \frac{F}{M} = \frac{G}{N}$$

hold is either spherical or plane.

Ex. 3. The centre of curvature at any point of a curve drawn on a surface is the projection upon its osculating plane of the centre of curvature of the normal section of the surface which touches the curve at the given point.

EXAMPLES IV

1. Taking x, y as parameters, calculate the fundamental magnitudes and the normal to the surface

$$2z = ax^2 + 2hxy + by^2.$$

2. For the surface of revolution

$$x = u \cos \phi, \quad y = u \sin \phi, \quad z = f(u),$$

with u, ϕ as parameters, show that

$$E = 1 + f'^2, \quad F = 0, \quad G = u^2, \quad H^2 = u^2(1 + f'^2),$$
$$\mathbf{n} = u(-f' \cos \phi, \; -f' \sin \phi, \; 1)/H,$$
$$L = \frac{uf''}{H}, \quad M = 0, \quad N = \frac{u^2 f'}{H}, \quad T^2 = \frac{u^3 f' f''}{H^2}.$$

3. Calculate the fundamental magnitudes and the unit normal for the conoid

$$x = u \cos \phi, \quad y = u \sin \phi, \quad z = f(\phi),$$

with u, ϕ as parameters.

4. On the surface generated by the binormals of a twisted curve, the position vector of the current point may be expressed $\mathbf{r} + u\mathbf{b}$ where $\mathbf{r}$ and $\mathbf{b}$ are functions of s. Taking u, s as parameters, show that

$$E = 1, \quad F = 0, \quad G = 1 + \tau^2 u^2, \quad H^2 = 1 + \tau^2 u^2,$$
$$\mathbf{n} = (\bar{\mathbf{n}} + \tau u \mathbf{t})/H,$$

where $\bar{\mathbf{n}}$ is the principal normal to the curve; also that

$$L = 0, \quad M = -\tau/H, \quad N = (\kappa + \kappa \tau^2 u^2 - \tau' u)/H.$$

5. When the equation of the surface is given in Monge's form $z = f(x, y)$, the coordinates x, y may be taken as parameters. If, as usual, p, q are the derivatives of z of the first order, and r, s, t those of the second order, show that

$$E = 1 + p^2, \quad F = pq, \quad G = 1 + q^2, \quad H^2 = 1 + p^2 + q^2,$$
$$\mathbf{n} = (-p, \; -q, \; 1)/H,$$
$$L = \frac{r}{H}, \quad M = \frac{s}{H}, \quad N = \frac{t}{H}, \quad T^2 = \frac{rt - s^2}{H^2}.$$

Deduce that T^2 is zero for a developable surface.

6. Find the tangent of the angle between the two directions on the surface determined by the quadratic

$$P\,du^2 + Q\,du\,dv + R\,dv^2 = 0.$$

Let du/dv and $\delta u/\delta v$ be the roots of this quadratic in du/dv. Then, by the first two results of Art. 24,

$$\tan\psi = \frac{\sin\psi}{\cos\psi} = \frac{H\left|\dfrac{du}{dv} - \dfrac{\delta u}{\delta v}\right|}{E\dfrac{du}{dv}\dfrac{\delta u}{\delta v} + F\left(\dfrac{du}{dv} + \dfrac{\delta u}{\delta v}\right) + G}$$

$$= \frac{H\sqrt{\dfrac{Q^2}{P^2} - \dfrac{4R}{P}}}{\dfrac{ER}{P} - \dfrac{FQ}{P} + G} = \frac{H\sqrt{Q^2 - 4PR}}{ER - FQ + GP}.$$

7. By means of the formulae of Art. 27 show that

$$T^2\mathbf{n}_1 \times \mathbf{n} = H\,(f\mathbf{n}_1 - e\mathbf{n}_2),$$
$$T^2\mathbf{n}_2 \times \mathbf{n} = H\,(g\mathbf{n}_1 - f\mathbf{n}_2),$$

where

$$H^2 e = EM^2 - 2FLM + GL^2,$$
$$H^2 f = EMN - F(LN + M^2) + GLM,$$
$$H^2 g = EN^2 - 2FMN + GM^2.$$

8. From a given point P on a surface a length PQ is laid off along the normal equal to twice the radius of normal curvature for a given direction through P, and a sphere is described on PQ as diameter. Any curve is drawn on the surface, passing through P in the given direction. Prove that its circle of curvature at P is the circle in which its osculating plane at P cuts the sphere.

9. Show that the curves $du^2 - (u^2 + c^2)\,d\phi^2 = 0$ form an orthogonal system on the helicoid of Ex. 1, Art. 26.

10. On the surface generated by the tangents to a twisted curve, find the differential equation of the curves which cut the generators at a constant angle β.

11. Find the equations of the surface of revolution for which

$$ds^2 = du^2 + (a^2 - u^2)\,dv^2.$$

12. Show that the curvature κ at any point P of the curve of intersection of two surfaces is given by

$$\kappa^2 \sin^2\theta = \kappa_1^2 + \kappa_2^2 - 2\kappa_1\kappa_2\cos\theta,$$

where κ_1, κ_2 are the normal curvatures of the surfaces in the direction of the curve at P, and θ is the angle between their normals at that point.

13. Prove the formulae (19) of Art. 27.

From (17) of that Art. it follows that

$$T^2\mathbf{n}_1 \times \mathbf{r}_1 = (EM - FL)\,\mathbf{n}_1 \times \mathbf{n}_2,$$

and therefore $\quad T^2\mathbf{n}\bullet\mathbf{n}_1\times\mathbf{r}_1=(EM-FL)\,\mathbf{n}\bullet\mathbf{n}_1\times\mathbf{n}_2$
$$=(EM-FL)\,T^2/H.$$
Thus $\qquad H[\mathbf{n},\,\mathbf{n}_1,\,\mathbf{r}_1]=EM-FL,$
and similarly for the others.

14. Prove the formulae
$$Hn\times\mathbf{n}_1=M\mathbf{r}_1-L\mathbf{r}_2\Big\}.$$
$$Hn\times\mathbf{n}_2=N\mathbf{r}_1-M\mathbf{r}_2\Big\}$$

On differentiating the formula $H\mathbf{n}=\mathbf{r}_1\times\mathbf{r}_2$ with respect to u we have
$$H_1\mathbf{n}+H\mathbf{n}_1=\mathbf{r}_1\times\mathbf{r}_{21}+\mathbf{r}_{11}\times\mathbf{r}_2.$$
Form the cross product of each side with $\mathbf{n}$. Then
$$H\mathbf{n}\times\mathbf{n}_1=\mathbf{n}\bullet\mathbf{r}_{21}\mathbf{r}_1-\mathbf{n}\bullet\mathbf{r}_{11}\mathbf{r}_2$$
$$=M\mathbf{r}_1-L\mathbf{r}_2,$$
and similarly for the other.

By substituting in these the values of $\mathbf{r}_1$ and $\mathbf{r}_2$ given in (17) we could deduce the formulae of Ex. 7.

15. Helicoids. A *helicoid* is the surface generated by a curve which is simultaneously rotated about a fixed axis and translated in the direction of the axis with a velocity proportional to the angular velocity of rotation. The plane sections through the axis are called *meridians*. No generality is lost by assuming the curve to be plane, and the surface can be generated by the helicoidal motion of a meridian.

Take the axis of rotation as z-axis. Let u be the perpendicular distance of a point from the axis, and v the inclination of the meridian plane through the point to the zx-plane. Then, if the meridian $v=0$ is given by $z=f(u)$, the coordinates of a current point on the surface are
$$x=u\cos v,\quad y=u\sin v,\quad z=f(u)+cv,$$
where c is constant, and $2\pi c$ is called the *pitch* of the helicoidal motion. From these it follows that
$$E=1+f_1^2,\quad F=cf_1,\quad G=u^2+c^2,\quad H^2=c^2+u^2(1+f_1^2).$$
The parametric curves are orthogonal only when c is zero or $f(u)$ constant. The former case is that of a surface of revolution. The latter is the case of a *right helicoid* which is generated by the helicoidal motion of a straight line cutting the axis at right angles (Art. 26, Ex. 1). The unit normal to the general helicoid is
$$\mathbf{n}=\frac{1}{H}(c\sin v-uf_1\cos v,\ -c\cos v-uf_1\sin v,\ u),$$
and the second order magnitudes are
$$L=\frac{uf_{11}}{H},\quad M=-\frac{c}{H},\quad N=\frac{u^2f_1}{H}.$$
The parametric curves $u=$ const. are obviously helices.

16. Find the curvature of a normal section of a helicoid.

17. The locus of the mid-points of the chords of a circular helix is a right helicoid.

CHAPTER IV

CURVES ON A SURFACE

Lines of Curvature

29. Principal directions. The normals at consecutive points of a surface do not in general intersect; but at any point P there are two directions on the surface, at right angles to each other, such that the normal at a consecutive point in either of these directions meets the normal at P. These are called the *principal directions* at P. To prove this property, let $\mathbf{r}$ be the position vector of P and $\mathbf{n}$ the unit normal there. Let $\mathbf{r} + d\mathbf{r}$ be a consecutive point in the direction du, dv, and $\mathbf{n} + d\mathbf{n}$ the unit normal at this point. The normals will intersect if $\mathbf{n}$, $\mathbf{n} + d\mathbf{n}$ and $d\mathbf{r}$ are coplanar, that is to say, if $\mathbf{n}$, $d\mathbf{n}$, $d\mathbf{r}$ are coplanar. This will be so if their scalar triple product vanishes, so that

$$[\mathbf{n}, d\mathbf{n}, d\mathbf{r}] = 0 \quad \dots \dots \dots \dots \dots \dots (1).$$

This condition may be expanded in terms of du, dv. For

$$d\mathbf{n} = \mathbf{n}_1 du + \mathbf{n}_2 dv,$$
$$d\mathbf{r} = \mathbf{r}_1 du + \mathbf{r}_2 dv,$$

and the substitution of these values in (1) gives

$$[\mathbf{n}, \mathbf{n}_1, \mathbf{r}_1]\, du^2 + \{[\mathbf{n}, \mathbf{n}_1, \mathbf{r}_2] + [\mathbf{n}, \mathbf{n}_2, \mathbf{r}_1]\}\, du\,dv + [\mathbf{n}, \mathbf{n}_2, \mathbf{r}_2]\, dv^2 = 0$$

which, by (19) of Art 27, is equivalent to

$$(EM - FL)\, du^2 + (EN - GL)\, du\,dv + (FN - GM)\, dv^2 = 0 \quad \dots (2).$$

This equation gives two values of the ratio $du : dv$, and therefore two directions on the surface for which the required property holds. And these two directions are at right angles, for they satisfy the condition of orthogonality (11) of Art. 24.

It follows from the above that, for displacement in a principal direction, $d\mathbf{n}$ is parallel to $d\mathbf{r}$. For $d\mathbf{r}$ is perpendicular to $\mathbf{n}$, and $d\mathbf{n}$ is also perpendicular to $\mathbf{n}$ since $\mathbf{n}$ is a unit vector. But these three vectors are coplanar, and therefore $d\mathbf{n}$ is parallel to $d\mathbf{r}$. Thus, for a principal direction, $\mathbf{n}'$ is parallel to $\mathbf{r}'$, the dash denoting arc-rate of change.

A curve drawn on the surface, and possessing the property that the normals to the surface at consecutive points intersect, is called a *line of curvature*. It follows from the above that the direction of a line of curvature at any point is a principal direction at that point. Through each point on the surface pass two lines of curvature cutting each other at right angles; and on the surface there are two systems of lines of curvature whose differential equation is (2). The point of intersection of consecutive normals along a line of curvature at P is called a *centre of curvature* of the surface; and its distance from P, measured in the direction of the unit normal **n**, is called a *(principal) radius of curvature* of the surface. The reciprocal of a principal radius of curvature is called a *principal curvature*. Thus at each point of the surface there are two principal curvatures, κ_a and κ_b, and these are the normal curvatures of the surface in the directions of the lines of curvature. They must not be confused with the curvatures of the lines of curvature. For the principal normal of a line of curvature is not in general the normal to the surface. In other words, the osculating plane of a line of curvature does not, as a rule, give a normal section of the surface; but the curvature of a line of curvature is connected with the corresponding principal curvature as in Meunier's Theorem.

The principal radii of curvature will be denoted by α, β. As these are the reciprocals of the principal curvatures, we have

$$\alpha\kappa_a = 1, \quad \beta\kappa_b = 1.$$

Those portions of the surface on which the two principal curvatures have the same sign are said to be *synclastic*. The surface of a sphere or of an ellipsoid is synclastic at all points. On the other hand if the principal curvatures have opposite signs on any part of the surface, this part is said to be *anticlastic*. The surface of a hyperbolic paraboloid is anticlastic at all points.

At any point of a surface there are two centres of curvature, one for each principal direction. Both lie on the normal to the surface, for they are the centres of curvature of normal sections tangential to the lines of curvature. The locus of the centres of curvature is a surface called the *surface of centres*, or the *centro-surface*. It consists of two branches, one corresponding to each system of lines of curvature. The properties of the centro-surface will be examined in a later chapter.

Ex. Joachimsthal's Theorem. *If the curve of intersection of two surfaces is a line of curvature on both, the surfaces cut at a constant angle. Conversely, if two surfaces cut at a constant angle, and the curve of intersection is a line of curvature on one of them, it is a line of curvature on the other also.*

Let $\mathbf{t}$ be the unit tangent to the curve of intersection, and $\mathbf{n}$, $\bar{\mathbf{n}}$ the unit normals at the same point to the two surfaces. Then $\mathbf{t}$ is perpendicular to $\mathbf{n}$ and $\bar{\mathbf{n}}$, and therefore parallel to $\mathbf{n} \times \bar{\mathbf{n}}$. Further, if the curve is a line of curvature on both surfaces, $\mathbf{t}$ is parallel to $\mathbf{n}'$ and $\bar{\mathbf{n}}'$, the dash as usual denoting arc-rate of change. Let θ be the inclination of the two normals. Then $\cos \theta = \mathbf{n} \cdot \bar{\mathbf{n}}$, and

$$\frac{d}{ds} \cos \theta = \mathbf{n}' \cdot \bar{\mathbf{n}} + \mathbf{n} \cdot \bar{\mathbf{n}}'.$$

But each of these terms vanishes because $\mathbf{n}'$ and $\bar{\mathbf{n}}'$ are both parallel to $\mathbf{t}$. Thus $\cos \theta$ is constant, and the surfaces cut at a constant angle.

Similarly if θ is constant, and the curve is a line of curvature on the first surface, all the terms of the above equation disappear except the last. Hence this must vanish also, showing that $\bar{\mathbf{n}}'$ is perpendicular to $\mathbf{n}$. But it is also perpendicular to $\bar{\mathbf{n}}$, because $\bar{\mathbf{n}}$ is a unit vector. Thus $\bar{\mathbf{n}}'$ is parallel to $\mathbf{n} \times \bar{\mathbf{n}}$, and therefore also to $\mathbf{t}$. The curve of intersection is thus a line of curvature on the second surface also.

30. First and second curvatures. To determine the principal curvatures at any point we may proceed as follows. Let $\mathbf{r}$ be the position vector of the point, $\mathbf{n}$ the unit normal and ρ a principal radius of curvature. Then the corresponding centre of curvature $\mathbf{s}$ is $\mathbf{r} + \rho \mathbf{n}$. For an infinitesimal displacement of the point along the line of curvature we have therefore

$$d\mathbf{s} = (d\mathbf{r} + \rho d\mathbf{n}) + \mathbf{n} d\rho.$$

The vector in brackets is tangential to the surface; and consequently, since $d\mathbf{s}$ has the direction of $\mathbf{n}$ (cf. Art. 74),

$$0 = d\mathbf{r} + \rho d\mathbf{n} \quad \dots\dots\dots\dots\dots(3),$$

or, if κ is the corresponding principal curvature,

$$0 = \kappa d\mathbf{r} + d\mathbf{n} \quad \dots\dots\dots\dots\dots(3').$$

This is the vector equivalent of *Rodrigues' formula*. It is of very great importance. Inserting the values of the differentials in terms of du and dv we may write it

$$(\kappa \mathbf{r}_1 + \mathbf{n}_1)\, du + (\kappa \mathbf{r}_2 + \mathbf{n}_2)\, dv = 0.$$

Forming the scalar products of this with $\mathbf{r}_1$ and $\mathbf{r}_2$ successively we have, by (15) of Art. 27,

$$\left.\begin{array}{l} (\kappa E - L)\, du + (\kappa F - M)\, dv = 0 \\ (\kappa F - M)\, du + (\kappa G - N)\, dv = 0 \end{array}\right\} \dots\dots\dots(4).$$

These two equations determine the principal curvatures and the directions of the lines of curvature. On eliminating du/dv we have for the principal curvatures

$$(\kappa E - L)(\kappa G - N) = (\kappa F - M)^2,$$

or $\quad H^2\kappa^2 - (EN - 2FM + GL)\kappa + T^2 = 0 \quad \ldots\ldots(5),$

a quadratic, giving two values of κ as required.

The *first curvature* of the surface at any point may be defined as the sum of the principal curvatures*. We will denote it by J. Thus

$$J = \kappa_a + \kappa_b.$$

Being the sum of the roots of the quadratic (5) it is given by

$$J = \frac{1}{H^2}(EN - 2FM + GL) \quad \ldots\ldots\ldots(6).$$

The *second curvature*, or *specific curvature*, of the surface at any point is the product of the principal curvatures. It is also called the *Gauss curvature*, and is denoted by K. It is equal to the product of the roots of (5), so that

$$K = \kappa_a\kappa_b = \frac{T^2}{H^2} \quad \ldots\ldots\ldots\ldots(7).$$

When the principal curvatures have been determined from (5), the directions of the lines of curvature are given by either of the equations (4). Thus corresponding to the principal curvature κ_a the principal direction is given by

$$\frac{du}{dv} = -\left(\frac{\kappa_a F - M}{\kappa_a E - L}\right) \text{ or } -\left(\frac{\kappa_a G - N}{\kappa_a F - M}\right),$$

and similarly for the other principal direction.

The directions of the lines of curvature may, of course, be found independently by eliminating κ from the equations (4). This leads to

$$(EM - FL)\,du^2 + (EN - GL)\,du\,dv + (FN - GM)\,dv^2 = 0\ldots(8),$$

the same equation as (2) found by a different method. It may be remarked that this is also the equation giving the directions of

---

* Some writers call J the *mean* curvature and K the *total* curvature. On this question see remarks in the Preface and also on p. 264, Note II.

maximum and minimum normal curvature at the point. For, the value of the normal curvature, being

$$\kappa_n = \frac{L\,du^2 + 2M\,du\,dv + N\,dv^2}{E\,du^2 + 2F\,du\,dv + G\,dv^2} \quad \dots\dots\dots\dots(9),$$

as found in Art. 28, is a function of the ratio $du:dv$; and if its derivative with respect to this ratio is equated to zero, we obtain the same equation (8) as before. Thus *the principal directions at a point are the directions of greatest and least normal curvature.*

The equation (8), however, fails to determine these directions when the coefficients vanish identically, that is to say when

$$E:F:G = L:M:N \dots\dots\dots\dots\dots(10).$$

In this case the normal curvature, as determined by (9), is independent of the ratio $du:dv$, and therefore has the same value for all directions through the point. Such a point is called an *umbilic* on the surface.

If the *amplitude* of normal curvature, A, and the *mean* normal curvature, B, are defined by

$$A = \tfrac{1}{2}(\kappa_b - \kappa_a), \quad B = \tfrac{1}{2}(\kappa_b + \kappa_a) \quad \dots\dots\dots\dots(11),$$

it follows that

$$\kappa_a = B - A, \quad \kappa_b = B + A \dots\dots\dots\dots(12).$$

Hence the second curvature may be expressed

$$K = B^2 - A^2.$$

We may also mention in passing that, when the first curvature vanishes at all points, the surface is called a *minimal surface*. The properties of such surfaces will be examined in a later chapter.

Ex. 1. Find the principal curvatures and the lines of curvature on the right helicoid

$$x = u \cos\phi, \quad y = u \sin\phi, \quad z = c\phi.$$

The fundamental magnitudes for this surface were found in Ex. 1, Art. 26. Their values are

$$E = 1, \quad F = 0, \quad G = u^2 + c^2, \quad H^2 = u^2 + c^2,$$

$$L = 0, \quad M = -\frac{c}{H}, \quad N = 0, \quad T^2 = -\frac{c^2}{H^2}.$$

The formula (5) for the principal curvatures then becomes

$$(u^2 + c^2)^2\,\kappa^2 - c^2 = 0,$$

whence

$$\kappa = \pm\frac{c}{u^2 + c^2}.$$

The first curvature is therefore zero, so that the surface is a minimal surface. The second curvature is

$$K = -\frac{c^2}{(u^2+c^2)^2}.$$

The differential equation (2) for the lines of curvature becomes

$$-cdu^2 + (u^2+c^2)\,cd\phi^2 = 0,$$

that is

$$d\phi = \pm\frac{du}{\sqrt{u^2+c^2}}.$$

Ex. 2. Find the principal directions and the principal curvatures on the surface

$$x = a\,(u+v), \quad y = b\,(u-v), \quad z = uv.$$

It was shown in Ex. 2, Art. 26, that

$$E = a^2+b^2+v^2, \quad F = a^2-b^2+uv, \quad G = a^2+b^2+u^2,$$
$$H^2 = 4a^2b^2 + a^2\,(u-v)^2 + b^2\,(u+v)^2,$$

and also

$$L = 0, \quad M = -\frac{2ab}{H}, \quad N = 0, \quad T^2 = -\frac{4a^2b^2}{H^2}.$$

The differential equation (2) for the lines of curvature therefore gives

$$(a^2+b^2+v^2)\,du^2 - (a^2+b^2+u^2)\,dv^2 = 0,$$

or

$$\frac{du}{\sqrt{a^2+b^2+u^2}} = \pm\frac{dv}{\sqrt{a^2+b^2+v^2}}.$$

The equation (5) for the principal curvatures becomes

$$H^4\kappa^2 - 4ab\,H\,(a^2-b^2+uv)\,\kappa - 4a^2b^2 = 0,$$

so that the specific curvature is $K = -\dfrac{4a^2b^2}{H^4}$,

and the first curvature is

$$J = 4ab\,(a^2-b^2+uv)/H^3.$$

Ex. 3. Find the principal curvatures etc. on the surface generated by the binormals of a twisted curve.

The position vector of the current point on the surface may be expressed

$$\mathbf{R} = \mathbf{r} + u\mathbf{b},$$

where $\mathbf{r}$ and $\mathbf{b}$ are functions of the arc-length s. Taking u, s as parameters, and using dashes as usual to denote s-derivatives of quantities belonging to the curve, we have

$$\mathbf{R}_1 = \mathbf{b}, \quad \mathbf{R}_2 = \mathbf{t} - u\tau\bar{\mathbf{n}},$$

where $\bar{\mathbf{n}}$ is the unit principal normal to the curve. Hence

$$E = 1, \quad F = 0, \quad G = 1 + \tau^2u^2, \quad H^2 = 1 + \tau^2u^2,$$

and the unit normal to the surface is

$$\mathbf{n} = \frac{\mathbf{R}_1 \times \mathbf{R}_2}{H} = \frac{\bar{\mathbf{n}} + \tau u\mathbf{t}}{H}.$$

Further

$$\mathbf{R}_{11} = 0, \quad \mathbf{R}_{12} = -\tau\bar{\mathbf{n}},$$
$$\mathbf{R}_{22} = (\kappa - u\tau')\,\bar{\mathbf{n}} + u\tau\,(\kappa\mathbf{t} - \tau\mathbf{b}),$$

and therefore $\qquad L=0, \qquad\qquad\qquad M=-\dfrac{\tau}{H}$,

$$N=(\kappa+\kappa\tau^2 u^2-\tau'u)/H, \quad T^2=-\dfrac{\tau^2}{H^2}.$$

The equation (5) for the principal radii of curvature then becomes

$$(1+\tau^2 u^2)^2-\sqrt{1+\tau^2 u^2}\,(\kappa+\kappa\tau^2 u^2-\tau'u)\,\rho-\tau^2\rho^2=0.$$

The Gauss curvature is therefore

$$K=-\dfrac{\tau^2}{(1+\tau^2 u^2)^2},$$

and the first curvature

$$J=\dfrac{\kappa+\kappa\tau^2 u^2-\tau'u}{(1+\tau^2 u^2)^{\frac{3}{2}}}.$$

For points on the given curve, $u=0$. At such points the Gauss curvature is $-\tau^2$, and the first curvature is κ.

The differential equation of the lines of curvature reduces to

$$\tau\,du^2-(\kappa+\kappa\tau^2 u^2-\tau'u)\,du\,ds-(1+\tau^2 u^2)\,\tau\,ds^2=0.$$

31. Euler's theorem. It is sometimes convenient to refer the surface to its lines of curvature as parametric curves. If this is done the differential equation (2) for the lines of curvature becomes identical with the differential equation of the parametric curves, that is

$$du\,dv=0.$$

Hence we must have

$$EM-FL=0, \quad FN-GM=0,$$

and $\qquad\qquad\qquad EN-GL\neq 0.$

From the first two relations it follows that

$$\left.\begin{array}{r}(EN-GL)\,M=0\\(EN-GL)\,F=0\end{array}\right\},$$

and therefore, since the coefficient of F and M does not vanish,

$$F=0, \quad M=0 \qquad\qquad\dots\dots\dots\dots\dots(13).$$

These are the necessary and sufficient conditions that the parametric curves be lines of curvature. The condition $F=0$ is that of orthogonality satisfied by all lines of curvature. The significance of the condition $M=0$ will appear shortly (Art. 36).

We may now prove Euler's theorem, expressing the normal curvature in any direction in terms of the principal curvatures at the point. Let the lines of curvature be taken as parametric curves, so that $F = M = 0$. The principal curvature κ_a, being the normal curvature for the direction $dv = 0$, is by (9)

$$\kappa_a = L/E,$$

and similarly the principal curvature for the direction $du = 0$ is

$$\kappa_b = N/G.$$

Consider a normal section of the surface in the direction du, dv, making an angle ψ with the principal direction $dv = 0$. Then by (7) of Art. 24 and Note I, since $F = 0$, we have

$$\cos \psi = \sqrt{E}\,\frac{du}{ds},$$

and

$$\sin \psi = \sqrt{G}\,\frac{dv}{ds}.$$

The curvature κ_n of this normal section is by (9)

$$\kappa_n = L \left(\frac{du}{ds}\right)^2 + N \left(\frac{dv}{ds}\right)^2$$

$$= \frac{L}{E}\cos^2 \psi + \frac{N}{G}\sin^2 \psi,$$

so that

$$\kappa_n = \kappa_a \cos^2 \psi + \kappa_b \sin^2 \psi \dots\dots\dots\dots(14).$$

This is *Euler's theorem* on normal curvature. An immediate and important consequence is the theorem, associated with the name of Dupin, that *the sum of the normal curvatures in two directions at right angles is constant, and equal to the sum of the principal curvatures.*

When the surface is anticlastic in the neighbourhood of the point considered, the principal curvatures have opposite signs, and the normal curvature therefore vanishes for the directions given by

$$\tan \psi = \pm \sqrt{-\kappa_a/\kappa_b}$$

$$= \pm \sqrt{-\frac{\beta}{\alpha}},$$

where α, β are the principal radii of curvature. But where the surface is synclastic, the curvature of any normal section has the same sign as the principal curvatures, that is to say, all normal sections are concave in the same direction. The surface in the neighbourhood of the point then lies entirely on one side of the tangent plane at the point. The same result may also be deduced from the expression

$$\tfrac{1}{2}\,(L\,du^2 + 2M\,du\,dv + N\,dv^2),$$

found in Art. 26 for the length p of the perpendicular on the tangent plane from a point near the point of contact. For if K is positive, $LN - M^2$ is positive by (7), and therefore the above expression for p never changes sign with variation of du/dv.

Ex. If B is the mean normal curvature and A the amplitude, deduce from Euler's theorem that

$$\kappa_n = B - A \cos 2\psi,$$
$$\kappa_n - \kappa_a = 2A \sin^2 \psi,$$
$$\kappa_b - \kappa_n = 2A \cos^2 \psi.$$

32. Dupin's indicatrix. Consider the section of the surface by a plane parallel and indefinitely close to the tangent plane at the point P. Suppose first that the surface is synclastic in the neighbourhood of P. Then near P it lies entirely on one side of the tangent plane. Let the plane be taken on this (concave) side of the surface, parallel to the tangent plane at P, and at an

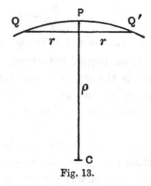

Fig. 13.

infinitesimal distance from it, whose measure is h in the direction of the unit normal **n**. Thus h has the same sign as the principal radii of curvature, α and β. Consider also any normal plane QPQ' through P, cutting the former plane in QQ'. Then if ρ is the radius of curvature of this normal section, and $2r$ the length of QQ', we have

$$r^2 = 2h\rho$$

to the first order. If ψ is the inclination of this normal section to the principal direction $dv = 0$, Euler's theorem gives

$$\frac{1}{\alpha} \cos^2 \psi + \frac{1}{\beta} \sin^2 \psi = \frac{1}{\rho} = \frac{2h}{r^2}.$$

If then we write $\xi = r \cos \psi$ and $\eta = r \sin \psi$ we have

$$\frac{\xi^2}{\alpha} + \frac{\eta^2}{\beta} = 2h.$$

Thus the section of the surface by the plane parallel to the tangent plane at P, and indefinitely close to it, is similar and similarly situated to the ellipse

$$\frac{\xi^2}{|\alpha|} + \frac{\eta^2}{|\beta|} = 1 \quad\dots\dots\dots\dots\dots(15),$$

whose axes are tangents to the lines of curvature at P. This ellipse is called the *indicatrix* at the point P, and P is said to be an *elliptic* point. It is sometimes described as a point of positive curvature, because the second curvature K is positive.

Next suppose that the Gauss curvature K is negative at P, so that the surface is anticlastic in the neighbourhood. The principal radii, α and β, have opposite signs, and the surface lies partly on one side and partly on the other side of the tangent plane at P. Two planes parallel to this tangent plane, one on either side, and equidistant from it, cut the surface in the conjugate hyperbolas

$$\frac{\xi^2}{\alpha} + \frac{\eta^2}{\beta} = \pm\, 2h.$$

These are similar and similarly situated to the conjugate hyperbolas

$$\frac{\xi^2}{\alpha} + \frac{\eta^2}{\beta} = \pm\, 1 \quad\dots\dots\dots\dots\dots(16),$$

which constitute the indicatrix at P. The point P is then called a *hyperbolic* point, or a point of negative curvature. The normal curvature is zero in the directions of the asymptotes.

When K is zero at the point P it is called a *parabolic* point. One of the principal curvatures is zero, and the indicatrix is a pair of parallel straight lines.

33. The surface $z = f(x, y)$. It frequently happens that the equation of the surface is given in Monge's form

$$z = f(x, y).$$

Let x, y be taken as parameters and, with the usual notation for partial derivatives of z, let

$$z_1 = p,\ z_2 = q;\ z_{11} = r,\ z_{12} = s,\ z_{22} = t.$$

Then, if $\mathbf{r}$ is the position vector of a current point on the surface

$$\mathbf{r}_1 = (1, 0, p),$$
$$\mathbf{r}_2 = (0, 1, q),$$

and therefore

$$E = 1 + p^2, \quad F = pq, \quad G = 1 + q^2, \quad H^2 = 1 + p^2 + q^2.$$

The inclination ω of the parametric curves is given by

$$\cos \omega = \frac{pq}{\sqrt{(1 + p^2)(1 + q^2)}}.$$

The unit normal to the surface is

$$\mathbf{n} = \frac{\mathbf{r}_1 \times \mathbf{r}_2}{H} = (- p, - q, 1)/H.$$

Further

$$\mathbf{r}_{11} = (0, 0, r),$$
$$\mathbf{r}_{12} = (0, 0, s),$$
$$\mathbf{r}_{22} = (0, 0, t),$$

so that the second order magnitudes are

$$L = \frac{r}{H}, \quad M = \frac{s}{H}, \quad N = \frac{t}{H}; \quad T^2 = \frac{rt - s^2}{H^2} = \frac{rt - s^2}{1 + p^2 + q^2}.$$

The specific curvature is therefore

$$K = \frac{T^2}{H^2} = \frac{rt - s^2}{(1 + p^2 + q^2)^2},$$

and the first curvature is

$$J = \frac{1}{H^3} \{r(1 + q^2) - 2pqs + t(1 + p^2)\}.$$

The equation (5) for the principal curvatures becomes

$$H^4 \kappa^2 - H \{r(1 + q^2) - 2pqs + t(1 + p^2)\} \kappa + (rt - s^2) = 0,$$

and the differential equation of the lines of curvature is

$$\{s(1 + p^2) - rpq\} \, dx^2 + \{t(1 + p^2) - r(1 + q^2)\} \, dx \, dy$$
$$+ \{tpq - s(1 + q^2)\} \, dy^2 = 0.$$

Since for a developable surface $rt - s^2$ is identically zero (Art. 17), it follows from the above value of K that *the second curvature vanishes at all points of a developable surface; and conversely, if the specific curvature is identically zero, the surface is a developable.*

Ex. 1. Find the equation for the principal curvatures, and the differential equation of the lines of curvature, for the surfaces

(i) $2z = \dfrac{x^2}{a} + \dfrac{y^2}{b}$,   (ii) $3z = ax^3 + by^3$,   (iii) $z = c \tan^{-1} \dfrac{y}{x}$.

Ex. 2. The indicatrix at every point of the helicoid

$$z = c \tan^{-1} \frac{y}{x}$$

is a rectangular hyperbola.

Ex. 3. The indicatrix at a point of the surface $z = f(x, y)$ is a rectangular hyperbola if

$$(1 + p^2)\, t - 2pqs + (1 + q^2)\, r = 0.$$

Ex. 4. At a point of intersection of the paraboloid $xy = cz$ with the hyperboloid $x^2 + y^2 - z^2 + c^2 = 0$ the principal radii of the paraboloid are

$$z^2(1 \pm \sqrt{2})/c.$$

34. Surface of revolution. A surface of revolution may be generated by the rotation of a plane curve about an axis in its plane. If this is taken as the axis of z, and u denotes perpendicular distance from it, the coordinates of a point on the surface may be expressed

$$x = u \cos \phi, \quad y = u \sin \phi, \quad z = f(u),$$

the longitude ϕ being the inclination of the axial plane through the given point to the zx-plane. The parametric curves $v = \text{con-}$ stant are the "meridian lines," or intersections of the surface by the axial planes; the curves $u = \text{constant}$ are the "parallels," or intersections of the surface by planes perpendicular to the axis.

With u, ϕ as parameters, and $\mathbf{r}$ the position vector of a current point on the surface, we have

$$\mathbf{r}_1 = (\cos \phi, \ \sin \phi, \ f_1),$$
$$\mathbf{r}_2 = (-u \sin \phi, \ u \cos \phi, \ 0).$$

The first order magnitudes are therefore

$$E = 1 + f_1^2, \quad F = 0, \quad G = u^2; \quad H^2 = u^2(1 + f_1^2).$$

Since $F = 0$ it follows that the parallels cut the meridians orthogonally. The unit normal to the surface is

$$\mathbf{n} = (-f_1 u \cos \phi, \ -f_1 u \sin \phi, \ u)/H.$$

Further
$$\mathbf{r}_{11} = (0, \ 0, \ f_{11}),$$
$$\mathbf{r}_{12} = (-\sin \phi, \ \cos \phi, \ 0),$$
$$\mathbf{r}_{22} = (-u \cos \phi, \ -u \sin \phi, \ 0),$$

so that the second order magnitudes are

$$L = u f_{11}/H, \quad M = 0, \quad N = u^2 f_1/H; \quad T^2 = u^3 f_1 f_{11}/H^2.$$

Since F and M both vanish identically, *the parametric curves are the lines of curvature.*

The equation for the principal curvatures reduces to

$$u (1 + f_1^2)^2 \kappa^2 - \sqrt{1 + f_1^2} \{u f_{11} + f_1 (1 + f_1^2)\} \kappa + f_1 f_{11} = 0,$$

the roots of which are

$$\kappa_a = \frac{f_{11}}{(1 + f_1^2)^{\frac{3}{2}}} = \frac{\dfrac{d^2 f}{du^2}}{\left\{1 + \left(\dfrac{df}{du}\right)^2\right\}^{\frac{3}{2}}},$$

and

$$\kappa_b = \frac{f_1}{u \sqrt{1 + f_1^2}} = \frac{\dfrac{df}{du}}{u \sqrt{1 + \left(\dfrac{df}{du}\right)^2}}.$$

The first of these is the curvature of the generating curve. The second is the reciprocal of the length of the normal intercepted between the curve and the axis of rotation. The Gauss curvature is given by

$$K = \frac{f_1 f_{11}}{u (1 + f_1^2)^2},$$

and the first curvature by

$$J = \frac{u f_{11} + f_1 (1 + f_1^2)}{u (1 + f_1^2)^{\frac{3}{2}}}.$$

Ex. 1. If the surface of revolution is a minimal surface,

$$u \frac{d^2 f}{du^2} + \frac{df}{du} \left\{1 + \left(\frac{df}{du}\right)^2\right\} = 0.$$

Hence show that the only real minimal surface of revolution is that formed by the revolution of a catenary about its directrix.

Ex. 2. On the surface formed by the revolution of a parabola about its directrix, one principal curvature is double the other.

EXAMPLES V

1. The moment about the origin of the unit normal $\mathbf{n}$ at a point $\mathbf{r}$ of the surface is $\mathbf{m} = \mathbf{r} \times \mathbf{n}$. Prove that the differential equation of the lines of curvature is

$$d\mathbf{m} \cdot d\mathbf{n} = 0.$$

2. Find equations for the principal radii, the lines of curvature, and the first and second curvatures of the following surfaces:

  (i) the conoid $x = u \cos \theta, \quad y = u \sin \theta, \quad z = f(\theta);$

  (ii) the catenoid

$$x = u \cos \theta, \quad y = u \sin \theta, \quad z = c \log (u + \sqrt{u^2 - c^2});$$

  (iii) the cylindroid $z (x^2 + y^2) = 2m xy;$

(iv) the surface $\qquad 2z = ax^2 + 2hxy + by^2$;

(v) the surface
$$x = 3u(1+v^2) - u^3, \quad y = 3v(1+u^2) - v^3, \quad z = 3(u^2 - v^2);$$

(vi) the surface
$$\frac{x}{a} = \frac{1+uv}{u+v}, \quad \frac{y}{b} = \frac{u-v}{u+v}, \quad \frac{z}{c} = \frac{1-uv}{u+v};$$

(vii) the surface $\qquad xyz = a^3.$

3. The lines of curvature of the paraboloid $xy = az$ lie on the surfaces
$$\sinh^{-1}\frac{x}{a} \pm \sinh^{-1}\frac{y}{a} = \text{const.}$$

4. Show that the surface
$$4a^2z^2 = (x^2 - 2a^2)(y^2 - 2a^2)$$
has a line of umbilics lying on the sphere
$$x^2 + y^2 + z^2 = 4a^2.$$

5. On the surface generated by the tangents to a twisted curve the current point is $\mathbf{R} = \mathbf{r} + u\mathbf{t}$. Taking u, s as parameters, prove that
$$E = 1, \quad F = 1, \quad G = 1 + u^2\kappa^2, \quad H^2 = u^2\kappa^2,$$
$$\mathbf{n} = \mathbf{b},$$
$$L = 0, \quad M = 0, \quad N = u\kappa\tau, \quad T^2 = 0,$$
$$K = 0, \quad J = \frac{\tau}{u\kappa}; \quad \kappa_a = 0, \quad \kappa_b = \frac{\tau}{u\kappa}.$$
The lines of curvature are
$$s = \text{const.}; \quad u + s = \text{const.}$$

6. Examine, as in Ex. 5, the curvature of the surface generated by the principal normals of a twisted curve.

7. Examine the curvature of the surface generated by the radii of spherical curvature of a twisted curve.

8. Show that the equation of the indicatrix, referred to the tangents to the parametric curves as (oblique) axes, is
$$\frac{L}{E}\xi^2 + \frac{2M}{\sqrt{EG}}\xi\eta + \frac{N}{G}\eta^2 = 1.$$

9. Calculate the first and second curvatures of the helicoid [Examples IV, (15)]
$$x = u\cos v, \quad y = u\sin v, \quad z = f(u) + cv,$$
and show that the latter is constant along a helix ($u = $ const.).

10. Show that the lines of curvature of the helicoid in Ex. 9 are given by
$$c(1 + f_1^2 + uf_1f_{11})\,du^2 + [(u^2 + c^2)\,uf_{11} - (1 + f_1^2)\,u^2f_1]\,du\,dv$$
$$- c(u^2 + c^2 + u^2f_1^2)\,dv^2 = 0.$$
The meridians will be lines of curvature if
$$1 + f_1^2 + uf_1f_{11} = 0.$$

11. Find the equations of the helicoid generated by a circle of radius a, whose plane passes through the axis; and determine the lines of curvature on the surface.

CONJUGATE SYSTEMS

35. Conjugate directions. Conjugate directions at a given point P on the surface may be defined as follows. Let Q be a point on the surface adjacent to P, and let PR be the line of intersection of the tangent planes at P and Q. Then, as Q tends to coincidence with P, the limiting directions of PQ and PR are said to be *conjugate directions* at P. Thus the characteristic of the tangent plane, as the point of contact moves along a given curve, is the tangent line in the direction conjugate to that of the curve at the point of contact. In other words *the tangent planes to the surface along a curve C envelop a developable surface each of whose generators has the direction conjugate to that of C at their point of intersection.*

To find an analytical expression of the condition that two directions may be conjugate, let $\mathbf{n}$ be the unit normal at P where the parameter values are u, v, and $\mathbf{n} + d\mathbf{n}$ that at Q where the values are $u + du$, $v + dv$. If R is the adjacent point to P, in the direction of the intersection of the tangent planes at P and Q, we may denote the vector PR by $\delta\mathbf{r}$ and the parameter values at R by $u + \delta u$, $v + \delta v$. Then since PR is parallel to the tangent planes at P and Q, $\delta\mathbf{r}$ is perpendicular both to $\mathbf{n}$ and to $\mathbf{n} + d\mathbf{n}$. Hence $\delta\mathbf{r}$ is perpendicular to $d\mathbf{n}$, so that

$$d\mathbf{n} \cdot \delta\mathbf{r} = 0,$$

and consequently

$$(\mathbf{n}_1 du + \mathbf{n}_2 dv) \cdot (\mathbf{r}_1 \delta u + \mathbf{r}_2 \delta v) = 0.$$

Expanding this product and remembering that (Art. 27)

$$\mathbf{n}_1 \cdot \mathbf{r}_1 = -L, \quad \mathbf{n}_1 \cdot \mathbf{r}_2 = \mathbf{n}_2 \cdot \mathbf{r}_1 = -M, \quad \mathbf{n}_2 \cdot \mathbf{r}_2 = -N,$$

we obtain the relation

$$L\,du\,\delta u + M\,(du\,\delta v + \delta u\,dv) + N\,dv\,\delta v = 0 \quad \ldots\ldots(17).$$

This is the necessary and sufficient condition that the direction $\delta u/\delta v$ be conjugate to the direction du/dv; and the symmetry of the relation shows that *the property is a reciprocal one.* Moreover the equation is linear in each of the ratios $du : dv$ and $\delta u : \delta v$; so that to a given direction there is one and only one conjugate direction.

The condition (17) that two directions be conjugate may be expressed

$$L\frac{du}{dv}\frac{\delta u}{\delta v} + M\left(\frac{du}{dv} + \frac{\delta u}{\delta v}\right) + N = 0 \quad \ldots\ldots\ldots(17').$$

Hence the two directions given by the equation

$$P\,du^2 + Q\,du\,dv + R\,dv^2 = 0$$

will be conjugate provided

$$L\left(\frac{R}{P}\right) + M\left(-\frac{Q}{P}\right) + N = 0,$$

that is

$$LR - MQ + NP = 0 \quad\ldots\ldots\ldots\ldots(18).$$

Now the parametric curves are given by

$$du\,dv = 0,$$

which corresponds to the values $P = R = 0$ and $Q = 1$. Hence the directions of the parametric curves will be conjugate provided $M = 0$. We have seen that this condition is satisfied when the lines of curvature are taken as parametric curves. Hence *the principal directions at a point of the surface are conjugate directions*.

Let the lines of curvature be taken as parametric curves, so that $F = 0$ and $M = 0$. The directions du/dv and $\delta u/\delta v$ are inclined to the curve $v = \text{const.}$ at angles θ, θ' such that (Art. 24)

$$\tan\theta = \sqrt{\frac{G}{E}\frac{dv}{du}}, \qquad \tan\theta' = \sqrt{\frac{G}{E}\frac{\delta v}{\delta u}}.$$

Substituting from these equations in (17), and remembering that $M = 0$, we see that the two directions will be conjugate provided

$$\tan\theta\,\tan\theta' = -\frac{L}{E}\frac{G}{N} = -\frac{\beta}{\alpha},$$

that is to say, provided they are *parallel to conjugate diameters of the indicatrix*.

36. Conjugate systems. Consider the family of curves

$$\phi\,(u, v) = \text{const.}$$

The direction $\delta u/\delta v$ of a curve at any point is given by

$$\phi_1\,\delta u + \phi_2\,\delta v = 0.$$

The conjugate direction du/dv, in virtue of (17), is then determined by

$$(L\phi_2 - M\phi_1)\,du + (M\phi_2 - N\phi_1)\,dv = 0 \quad\ldots\ldots\ldots(19).$$

This is a differential equation of the first order and first degree, and therefore defines a one-parameter family of curves $\psi\,(u,v) = \text{const.}$ This and the family $\phi\,(u, v) = \text{const.}$ are said to form a *conjugate*

w.   6 *system.* At a point of intersection of two curves, one from each family, their directions are conjugate.

Further, given two families of curves

$$\phi(u, v) = \text{const.,}$$
$$\psi(u, v) = \text{const.,}$$

we may determine the condition that they form a conjugate system. For, the directions of the two curves through a point u, v are given by

$$\left.\begin{array}{l} \phi_1 \delta u + \phi_2 \delta v = 0 \\ \psi_1 du + \psi_2 dv = 0 \end{array}\right\}.$$

It then follows from (17') that these directions will be conjugate if

$$L\phi_2\psi_2 - M(\phi_1\psi_2 + \phi_2\psi_1) + N\phi_1\psi_1 = 0 \quad \ldots\ldots(20).$$

This is the necessary and sufficient condition that the two families of curves form a conjugate system. In particular the parametric curves $v = $ const., $u = $ const. will form a conjugate system if $M = 0$. This agrees with the result found in the previous Art. Thus $M = 0$ *is the necessary and sufficient condition that the parametric curves form a conjugate system.*

We have seen that when the lines of curvature are taken as parametric curves, both $F = 0$ and $M = 0$ are satisfied. Thus *the lines of curvature form an orthogonal conjugate system.* And they are the only orthogonal conjugate system. For, if such a system of curves exists, and we take them for parametric curves, then $F = 0$ and $M = 0$. But this shows that the parametric curves are then lines of curvature. Hence the theorem.

Ex. 1. The parametric curves are conjugate on the following surfaces:

(i) a surface of revolution

$$x = u \cos\phi, \quad y = u \sin\phi, \quad z = f(u);$$

(ii) the surface generated by the tangents to a curve, on which

$$\mathbf{R} = \mathbf{r} + u\mathbf{t}, \quad (u, s \text{ parameters});$$

(iii) the surface $x = \phi(u), \quad y = \psi(v), \quad z = f(u) + F(v);$

(iv) the surface $z = f(x) + F(y)$, where x, y are parameters;

(v) $x = A(u-a)^m(v-a)^n, \quad y = B(u-b)^m(v-b)^n, \quad z = C(u-c)^m(v-c)^n,$ where A, B, C, a, b, c are constants.

Ex. 2. Prove that, at any point of the surface, the sum of the radii of normal curvature in conjugate directions is constant.

ASYMPTOTIC LINES

37. Asymptotic lines. The asymptotic directions at a point
on the surface are the *self-conjugate* directions; and an asymptotic
line is a curve whose direction at every point is self-conjugate.
Consequently, if in equation (17) connecting conjugate directions
we put $\delta u/\delta v$ equal to du/dv, we obtain the differential equation of
the asymptotic lines on the surface

$$L\,du^2 + 2M\,du\,dv + N\,dv^2 = 0 \quad \ldots\ldots\ldots\ldots(21).$$

Thus there are two asymptotic directions at a point. They are real
and different when $M^2 - LN$ is positive, that is to say when the
specific curvature is negative. They are imaginary when K is
positive. They are identical when K is zero. In the last case the
surface is a developable, and the single asymptotic line through a
point is the generator.

Since the normal curvature in any direction is equal to

$$L u'^2 + 2M u'v' + N v'^2,$$

it vanishes for the asymptotic directions. These directions are
therefore the directions of the asymptotes of the indicatrix; hence
the name. They are at right angles when the indicatrix is a rectangular hyperbola, that is when the principal curvatures are equal
and opposite. Thus the asymptotic lines are orthogonal when the
surface is a minimal surface.

*The osculating plane at any point of an asymptotic line is the
tangent plane to the surface.* This may be proved as follows. Since
the tangent **t** to the asymptotic line is perpendicular to the normal
n to the surface, $\mathbf{n} \cdot \mathbf{t} = 0$. On differentiating this with respect to
the arc-length of the line, we have

$$\mathbf{n'} \cdot \mathbf{t} + \mathbf{n} \cdot (\kappa \bar{\mathbf{n}}) = 0,$$

where $\bar{\mathbf{n}}$ is the principal normal to the curve. Now the first term
in this equation vanishes, because, by Art. 35, **t** is perpendicular
to the rate of change of the unit normal in the conjugate direction,
and an asymptotic direction is self-conjugate. Thus $\mathbf{n'} \cdot \mathbf{t} = 0$ and
the last equation becomes

$$\mathbf{n} \cdot \bar{\mathbf{n}} = 0.$$

Then since both **t** and $\bar{\mathbf{n}}$ are perpendicular to the normal, the
osculating plane of the curve is tangential to the surface. The

binormal is therefore normal to the surface; and we may take its direction so that

$$b = n \quad \dots\dots\dots\dots\dots\dots(22).$$

Then the principal normal $\bar{n}$ is given by

$$\bar{n} = n \times t.$$

If the parametric curves be asymptotic lines, the differential equation (21) is identical with the differential equation of the parametric curves

$$du\,dv = 0.$$

Hence *the necessary and sufficient conditions that the parametric curves be asymptotic lines are*

$$L = 0, \quad N = 0, \quad M \neq 0.$$

In this case the differential equation of the lines of curvature becomes

$$E\,du^2 - G\,dv^2 = 0,$$

and the equation for the principal curvatures is

$$H^2 \kappa^2 + 2FM\kappa - M^2 = 0,$$

so that
$$K = -\frac{M^2}{H^2}, \qquad J = -\frac{2FM}{H^2} \quad \dots\dots\dots\dots(23).$$

38. Curvature and torsion. We have seen that the unit binormal to an asymptotic line is the unit normal to the surface; or $b = n$. The *torsion* τ is found by differentiating this relation with respect to the arc-length s, thus obtaining

$$-\tau\bar{n} = n',$$

where $\bar{n} = n \times r'$ is the principal normal to the curve. Forming the scalar product of each side with $\bar{n}$, we have

$$-\tau = n \times r' \cdot n',$$

so that
$$\tau = [n, n', r'] \quad \dots\dots\dots\dots\dots(24),$$

which is one formula for the torsion.

The scalar triple product in this formula is of the same form as that occurring in (1) Art. 29, the vanishing of which gave the differential equation of the lines of curvature. The expression (24) may then be expanded exactly as in Art. 29, giving for the torsion of an asymptotic line

$$\tau = \frac{1}{H}\{(EM - FL)\,u'^2 + (EN - GL)\,u'v' + (FN - GM)\,v'^2\}.$$

Suppose now that the asymptotic lines are taken as parametric curves. Then $L = N = 0$, and this formula becomes

$$\tau = \frac{M}{H} (Eu'^2 - Gv'^2).$$

Hence for the asymptotic line $dv = 0$ we have

$$\tau = \frac{M}{H} E \left(\frac{du}{ds}\right)^2 = \frac{M}{H} = \sqrt{-K} \quad \ldots\ldots\ldots\ldots(25)$$

in virtue of (23). Similarly for the asymptotic line $du = 0$ the torsion is

$$\tau = -\frac{M}{H} G \left(\frac{dv}{ds}\right)^2 = -\frac{M}{H} = -\sqrt{-K} \quad \ldots\ldots\ldots(25').$$

Thus *the torsions of the two asymptotic lines through a point are equal in magnitude and opposite in sign; and the square of either is the negative of the specific curvature.* This theorem is due to Beltrami and Enneper.

To find the *curvature* κ of an asymptotic line, differentiate the unit tangent $\mathbf{t} = \mathbf{r}'$ with respect to the arc-length s. Then

$$\kappa \bar{\mathbf{n}} = \mathbf{r}''.$$

Forming the scalar product of each side with the unit vector $\bar{\mathbf{n}} = \mathbf{n} \times \mathbf{r}'$, we have the result

$$\kappa = [\mathbf{n}, \mathbf{r}', \mathbf{r}''] \quad \ldots\ldots\ldots\ldots\ldots\ldots\ldots(26).$$

Ex. 1. On the surfaces in Ex. 1 and Ex. 2 of Art. 26 the parametric curves are asymptotic lines.

Ex. 2. On the surface $z = f(x, y)$ the asymptotic lines are

$$r\,dx^2 + 2s\,dx\,dy + t\,dy^2 = 0,$$

and their torsions are $\quad \pm \sqrt{s^2 - rt}/(1 + p^2 + q^2)$.

Ex. 3. On the surface of revolution (Art. 34) the asymptotic lines are

$$f_{11}\,du^2 + uf_1\,d\phi^2 = 0.$$

Write down the value of their torsions.

Ex. 4. Find the asymptotic lines, and their torsions, on the surface generated by the binormals to a twisted curve (Ex. 3, Art. 30).

Ex. 5. Find the asymptotic lines on the surface $z = y \sin x$.

ISOMETRIC LINES

39. Isometric parameters. Suppose that, in terms of the parameters u, v, the square of the linear element of the surface has the form $\quad ds^2 = \lambda\,(du^2 + dv^2) \quad \ldots\ldots\ldots\ldots\ldots(27),$

where λ is a function of u, v or a constant. Then the parametric curves are *orthogonal* because $F = 0$. Further, the lengths of elements of the parametric curves are $\sqrt{\lambda}\,du$ and $\sqrt{\lambda}\,dv$, and these are equal if $du = dv$. Thus the parametric curves corresponding to the values $u, u + du, v, v + dv$ bound a small square provided $du = dv$. In this way the surface may be mapped out into small squares by means of the parametric curves, the sides of any one square corresponding to equal increments in u and v.

More generally, if the square of the linear element has the form

$$ds^2 = \lambda\,(U\,du^2 + V\,dv^2) \quad\dots\dots\dots\dots\dots(28),$$

where U is a function of u only and V a function of v only, we may change the parameters to ϕ, ψ by the transformation

$$d\phi = \sqrt{U}\,du, \quad d\psi = \sqrt{V}\,dv.$$

This does not alter the parametric curves; for the curves $u = $ const. are identical with the curves $\phi = $ const.; and similarly the curves $v = $ const. are also the curves $\psi = $ const. The equation (28) then becomes

$$ds^2 = \lambda\,(d\phi^2 + d\psi^2) \quad\dots\dots\dots\dots\dots(29),$$

which is of the same form as (27). Whenever the square of the linear element has the form (28) so that, without alteration of the parametric curves, it may be reduced to the form (27), the parametric curves are called *isometric lines*, and the parameters *isometric parameters*. Sometimes the term *isothermal* or *isothermic* is used.

In the form (27) the fundamental magnitudes E and G are equal; but in the more general form (28) they are such that

$$\frac{E}{G} = \frac{U}{V} \quad\dots\dots\dots\dots\dots(30),$$

and therefore

$$\frac{\partial^2}{\partial u\partial v}\log\frac{E}{G} = 0\dots\dots\dots\dots\dots(31).$$

Either of these equations, in conjunction with $F = 0$, expresses *the condition that the parametric variables may be isometric.* For, if it is satisfied, ds^2 has the form (28) and may therefore be reduced to the form (27).

A simple example of isometric curves is afforded by the meridians and parallels on a *surface of revolution*. With the usual notation (Art. 34) $\quad x = u\cos\phi, \quad y = u\sin\phi, \quad z = f(u),$

since $\qquad E = 1 + f_1{}^2, \quad F = 0, \quad G = u^2,$

we have $\qquad ds^2 = (1 + f_1{}^2)\, du^2 + u^2\, d\phi^2$

$$= u^2 \left(\frac{1 + f_1{}^2}{u^2}\, du^2 + d\phi^2 \right) \quad \dots\dots\dots(32),$$

which is of the form (28). The parametric curves are the meridians $\phi = $ const. and the parallels $u = $ const. If we make the transformation

$$d\psi = \frac{1}{u} \sqrt{1 + f_1{}^2}\, du,$$

the curves $\psi = $ const. are the same as the parallels, and the square of the linear element becomes

$$ds^2 = u^2 \, (d\psi^2 + d\phi^2),$$

which is of the form (27). Thus *the meridians and the parallels of a surface of revolution are isometric lines.*

Ex. 1. Show that a system of confocal ellipses and hyperbolas are isometric lines in the plane.

Ex. 2. Determine $f(v)$ so that on the right conoid

$$x = u \cos v, \quad y = u \sin v, \quad z = f(v),$$

the parametric curves may be isometric lines.

Ex. 3. Find the surface of revolution for which

$$ds^2 = du^2 + (a^2 - u^2)\, dv^2.$$

NULL LINES

40. Null lines. The null lines (or *minimal curves*) on a surface are defined as the curves of zero length. They are therefore imaginary on a real surface, and their importance is chiefly analytic. The differential equation of the null lines is obtained by equating to zero the square of the linear element. It is therefore

$$E\,du^2 + 2F\,du\,dv + G\,dv^2 = 0 \quad \dots\dots\dots(33).$$

If the parametric curves are null lines, this equation must be equivalent to $du\,dv = 0$. Hence $E = 0$, $G = 0$ and $F \neq 0$. These are *the necessary and sufficient conditions that the parametric curves be null lines*. In this case the square of the linear element has the form

$$ds^2 = \lambda\, du\, dv,$$

where λ is a function of u, v or a constant; and the parameters u, v are then said to be *symmetric*.

When the parametric curves are null lines, so that

$$E = 0, \quad G = 0, \quad H^2 = -F^2,$$

the differential equation of the lines of curvature is

$$L\,du^2 - N\,dv^2 = 0,$$

the Gauss curvature is

$$K = \frac{LN - M^2}{-F^2},$$

and the first curvature

$$J = \frac{2M}{F}.$$

In the following pages our concern will be mainly with real curves and real surfaces. Only occasional reference will be made to null lines.

EXAMPLES VI

1. Find the asymptotic lines of the conoid

$$x = u \cos v, \quad y = u \sin v, \quad z = f(v),$$

and those of the cylindroid

$$x = u \cos v, \quad y = u \sin v, \quad z = m \sin 2v.$$

2. On the surface

$$x = 3u(1+v^2) - u^3, \quad y = 3v(1+u^2) - v^3, \quad z = 3(u^2 - v^2),$$

the asymptotic lines are $u \pm v = \text{const.}$

3. On the paraboloid $\quad 2z = \dfrac{x^2}{a^2} - \dfrac{y^2}{b^2},$

the asymptotic lines are $\quad \dfrac{x}{a} \pm \dfrac{y}{b} = \text{const.}$

4. Find the lines of curvature and the principal curvatures on the cylindroid

$$z(x^2 + y^2) = 2m\,xy.$$

5. If a plane cuts a surface everywhere at the same angle, the section is a line of curvature on the surface.

6. Along a line of curvature of a conicoid, one principal radius varies as the cube of the other.

7. Find the principal curvatures and the lines of curvature on the surface

$$z^2(x^2 + y^2) = c^4.$$

8. Find the asymptotic lines and the lines of curvature on the catenoid of revolution

$$u = c \cosh \frac{z}{c}.$$

9. If $a>b>c$, the ellipsoid $\dfrac{x^2}{a^2}+\dfrac{y^2}{b^2}+\dfrac{z^2}{c^2}=1$ has umbilici at the points

$$y=0,\quad x^2=\frac{a^2\,(a^2-b^2)}{a^2-c^2},\quad z^2=\frac{c^2\,(b^2-c^2)}{a^2-c^2}.$$

10. The only developable surfaces which have isometric lines of curvature are either conical or cylindrical.

11. Taking the asymptotic lines as parametric curves, and evaluating $[\mathbf{n},\ \mathbf{n}',\ \mathbf{r}']$ along the directions $v=$const. and $u=$const., verify the values $\pm\sqrt{-K}$ for the torsions of the asymptotic lines.

12. Show that the meridians and parallels on a sphere form an isometric system, and determine the isometric parameters.

13. Find the asymptotic lines on the surface

$$x=a\,(1+\cos u)\cot v,\quad y=a\,(1+\cos u),\quad z=\frac{a\cos u}{\sin v}.$$

14. Prove that the product of the radii of normal curvature in conjugate directions is a minimum for lines of curvature.

15. A curve, which touches an asymptotic line at P, and whose osculating plane is not tangential to the surface at P, has P for a point of inflection.

16. The normal curvature in a direction perpendicular to an asymptotic line is twice the mean normal curvature.

17. Show that the umbilici of the surface

$$\left(\frac{x}{a}\right)^{\frac{2}{3}}+\left(\frac{y}{b}\right)^{\frac{2}{3}}+\left(\frac{z}{c}\right)^{\frac{2}{3}}=1$$

lie on a sphere.

18. Examine the curvature, and find the lines of curvature, on the surface $xyz=abc$.

19. Show that the curvature of an asymptotic line, as given in (26) of Art. 38, may be expressed

$$(\mathbf{r}_1\bullet\mathbf{r}'\ \mathbf{r}_2\bullet\mathbf{r}''-\mathbf{r}_2\bullet\mathbf{r}'\ \mathbf{r}_1\bullet\mathbf{r}'')/H.$$

20. The asymptotic lines on the helicoid of Examples IV (15) are given by

$$uf_{11}\,du^2-2c\,du\,dv+u^2f_1\,dv^2=0.$$

CHAPTER V

THE EQUATIONS OF GAUSS AND OF CODAZZI

41. Gauss's formulae for $\mathbf{r}_{11}, \mathbf{r}_{12}, \mathbf{r}_{22}$. The second derivatives of $\mathbf{r}$ with respect to the parameters may be expressed in terms of $\mathbf{n}, \mathbf{r}_1$ and $\mathbf{r}_2$. Remembering that L, M, N are the resolved parts of $\mathbf{r}_{11}, \mathbf{r}_{12}, \mathbf{r}_{22}$ normal to the surface, we may write

$$\left. \begin{aligned} \mathbf{r}_{11} &= L\mathbf{n} + l\mathbf{r}_1 + \lambda\mathbf{r}_2 \\ \mathbf{r}_{12} &= M\mathbf{n} + m\mathbf{r}_1 + \mu\mathbf{r}_2 \\ \mathbf{r}_{22} &= N\mathbf{n} + n\mathbf{r}_1 + \nu\mathbf{r}_2 \end{aligned} \right\} \dots\dots\dots\dots\dots(1),$$

and the values of the coefficients $l, m, n, \lambda, \mu, \nu$ may be found as follows. Since

$$\mathbf{r}_1 \bullet \mathbf{r}_{11} = \frac{1}{2}\frac{\partial}{\partial u}\mathbf{r}_1^2 = \frac{1}{2}E_1,$$

and

$$\mathbf{r}_2 \bullet \mathbf{r}_{11} = \frac{\partial}{\partial u}(\mathbf{r}_1 \bullet \mathbf{r}_2) - \frac{1}{2}\frac{\partial}{\partial v}\mathbf{r}_1^2 = F_1 - \frac{1}{2}E_2,$$

we find from the first of (1), on forming the scalar product of each side with $\mathbf{r}_1$ and $\mathbf{r}_2$ successively,

$$\left. \begin{aligned} \tfrac{1}{2}E_1 &= lE + \lambda F \\ F_1 - \tfrac{1}{2}E_2 &= lF + \lambda G \end{aligned} \right\}.$$

Solving these for l and λ we have

$$\left. \begin{aligned} l &= \frac{1}{2H^2}(GE_1 - 2FF_1 + FE_2) \\ \lambda &= \frac{1}{2H^2}(2EF_1 - EE_2 - FE_1) \end{aligned} \right\} \dots\dots\dots\dots\dots(2).$$

Again since $\mathbf{r}_1 \bullet \mathbf{r}_{12} = \frac{1}{2}E_2$ and $\mathbf{r}_2 \bullet \mathbf{r}_{12} = \frac{1}{2}G_1$, we find from the second of (1), on forming the scalar product of each side with $\mathbf{r}_1$ and $\mathbf{r}_2$ successively,

$$\left. \begin{aligned} \tfrac{1}{2}E_2 &= mE + \mu F \\ \tfrac{1}{2}G_1 &= mF + \mu G \end{aligned} \right\}.$$

Solving these for m and μ we have

$$\left. \begin{aligned} m &= \frac{1}{2H^2}(GE_2 - FG_1) \\ \mu &= \frac{1}{2H^2}(EG_1 - FE_2) \end{aligned} \right\} \dots\dots\dots\dots\dots(3).$$

Similarly, using the relations $\mathbf{r}_1 \cdot \mathbf{r}_{22} = F_2 - \frac{1}{2}G_1$ and $\mathbf{r}_2 \cdot \mathbf{r}_{22} = \frac{1}{2}G_2$, we find from the third of (1)

$$n = \frac{1}{2H^2}(2GF_2 - GG_1 - FG_2) \\ \nu = \frac{1}{2H^2}(EG_2 - 2FF_2 + FG_1)$$ ....(4).

The formulae (1), with the values of the coefficients* given by (2), (3) and (4), are the equivalent of *Gauss's formulae* for $\mathbf{r}_{11}$, $\mathbf{r}_{12}$, $\mathbf{r}_{22}$, and may be referred to under this name.

When the parametric curves are *orthogonal*, the values of the above coefficients are greatly simplified. For, in this case, $F = 0$ and $H^2 = EG$, so that

$$\mathbf{r}_{11} = L\mathbf{n} + \frac{E_1}{2E}\mathbf{r}_1 - \frac{E_2}{2G}\mathbf{r}_2 \\ \mathbf{r}_{12} = M\mathbf{n} + \frac{E_2}{2E}\mathbf{r}_1 + \frac{G_1}{2G}\mathbf{r}_2 \\ \mathbf{r}_{22} = N\mathbf{n} - \frac{G_1}{2E}\mathbf{r}_1 + \frac{G_2}{2G}\mathbf{r}_2$$ ....(A).

If $\mathbf{a}$, $\mathbf{b}$ are *unit* vectors parallel to $\mathbf{r}_1$ and $\mathbf{r}_2$, we have

$$\mathbf{a} = \frac{\mathbf{r}_1}{\sqrt{E}}, \quad \mathbf{b} = \frac{\mathbf{r}_2}{\sqrt{G}},$$

while $\mathbf{a}$, $\mathbf{b}$, $\mathbf{n}$ form a right-handed system of unit vectors, mutually perpendicular. From these formulae we deduce immediately that

$$\frac{\partial \mathbf{a}}{\partial u} = \frac{L}{\sqrt{E}}\mathbf{n} - \frac{E_2}{2H}\mathbf{b} \\ \frac{\partial \mathbf{a}}{\partial v} = \frac{M}{\sqrt{E}}\mathbf{n} + \frac{G_1}{2H}\mathbf{b} \\ \frac{\partial \mathbf{b}}{\partial u} = \frac{M}{\sqrt{G}}\mathbf{n} + \frac{E_2}{2H}\mathbf{a} \\ \frac{\partial \mathbf{b}}{\partial v} = \frac{N}{\sqrt{G}}\mathbf{n} - \frac{G_1}{2H}\mathbf{a}$$ ....(B).

The derivatives of $\mathbf{a}$ are perpendicular to $\mathbf{a}$, and the derivatives of $\mathbf{b}$ are perpendicular to $\mathbf{b}$, since $\mathbf{a}$ and $\mathbf{b}$ are vectors of constant (unit) length.

* We refrain from introducing the Christoffel three-index symbols, having little occasion in the following pages to use the functions they represent.

Ex. 1. Show that for the surface $z = f(x, y)$, with x, y as parameters (Art. 33),

$$l = \frac{pr}{H^2}, \quad m = \frac{ps}{H^2}, \quad n = \frac{pt}{H^2},$$

$$\lambda = \frac{qr}{H^2}, \quad \mu = \frac{qs}{H^2}, \quad \nu = \frac{qt}{H^2}.$$

Ex. 2. For the surface of revolution (Art. 34) show that

$$l = u^2 f_1 f_{11}/H^2, \quad m = 0, \quad n = -u^3/H^2,$$

$$\lambda = 0, \qquad \mu = \frac{1}{u}, \quad \nu = 0.$$

Ex. 3. For the right helicoid (Ex. 1, Art. 26) prove that

$$l = 0, \quad m = 0, \qquad n = -u,$$

$$\lambda = 0, \quad \mu = u/(u^2 + c^2), \quad \nu = 0.$$

Ex. 4. For a surface whose linear element is given by

$$ds^2 = du^2 + D^2\, dv^2,$$

show that

$$l = 0, \quad m = 0, \qquad n = -DD_1,$$

$$\lambda = 0, \quad \mu = D_1/D, \quad \nu = D_2/D.$$

Ex. 5. *Liouville surfaces* are such that

$$ds^2 = (U + V)(P\, du^2 + Q\, dv^2),$$

where U, P are functions of u alone, and V, Q are functions of v alone. Prove that, for these surfaces,

$$l = \frac{1}{2}\left(\frac{U'}{U+V} + \frac{P'}{P}\right), \quad m = \frac{V'}{2(U+V)}, \quad n = -\frac{QU'}{2P(U+V)},$$

$$\lambda = -\frac{PV'}{2Q(U+V)}, \qquad \mu = \frac{U'}{2(U+V)}, \quad \nu = \frac{1}{2}\left(\frac{V'}{U+V} + \frac{Q'}{Q}\right).$$

Ex. 6. For the surface generated by the tangents to a twisted curve [Examples V (5)] show that

$$l = 0, \quad m = -\frac{1}{u}, \quad n = -\{(1 + u^2\kappa^2)\kappa + u\kappa'\}/u\kappa,$$

$$\lambda = 0, \quad \mu = \frac{1}{u}, \qquad \nu = (u\kappa' + \kappa)/u\kappa.$$

Ex. 7. For the surface generated by the binormals to a twisted curve (Ex. 3, Art. 30) show that

$$l = 0, \quad m = 0, \qquad n = -u\tau^2,$$

$$\lambda = 0, \quad \mu = \frac{u\tau^2}{1 + u^2\tau^2}, \quad \nu = \frac{u^2\tau\tau'}{1 + u^2\tau^2}.$$

Ex. 8. If the asymptotic lines are taken as parametric curves, prove that the curvature of the line $v = $ const. is $\lambda H/E^{\frac{3}{2}}$, and that of the line $u = $ const. is $-nH/G^{\frac{3}{2}}$.

Using the formula (26) of Art. 38 we have along the line $v=$const.

$$\mathbf{r}'=\frac{1}{\sqrt{E}}\mathbf{r}_1, \quad \mathbf{r}''=\frac{\mathbf{r}_{11}}{E}+\frac{\mathbf{r}_1}{\sqrt{E}}\frac{\partial}{\partial u}\left(\frac{1}{\sqrt{E}}\right),$$

and therefore $\qquad \mathbf{r}'\times\mathbf{r}''=\mathbf{r}_1\times\mathbf{r}_{11}/E^{\frac{3}{2}}.$

The curvature of the line is then

$$[\mathbf{n},\mathbf{r}',\mathbf{r}'']=[\mathbf{n},\mathbf{r}_1,\mathbf{r}_{11}]/E^{\frac{3}{2}}=[\mathbf{n},\mathbf{r}_1,\,l\mathbf{r}_1+\lambda\mathbf{r}_2]/E^{\frac{3}{2}}$$
$$=\lambda H/E^{\frac{3}{2}},$$

and similarly for the asymptotic line $u=$const.

Ex. 9. For a surface given by $ds^2=\phi\,(du^2+dv^2)$ show that

$$l=\tfrac{1}{2}\phi_1/\phi, \quad m=\tfrac{1}{2}\phi_2/\phi, \quad n=-\tfrac{1}{2}\phi_1/\phi,$$
$$\lambda=-\tfrac{1}{2}\phi_2/\phi, \quad \mu=\tfrac{1}{2}\phi_1/\phi, \quad \nu=\tfrac{1}{2}\phi_2/\phi.$$

Ex. 10. If the null lines are taken as parametric curves, show that

$$l=F_1/F, \quad m=0, \quad n=0,$$
$$\lambda=0, \quad \mu=0, \quad \nu=F_2/F.$$

42. Gauss characteristic equation.

The six fundamental magnitudes E, F, G, L, M, N are not functionally independent, but are connected by three differential relations. One of these, due to Gauss, is an expression for $LN-M^2$ in terms of E, F, G and their derivatives of the first two orders. It may be deduced from the formulae of the preceding Art. For, in virtue of these,

$$\mathbf{r}_{11}\cdot\mathbf{r}_{22}=LN+lnE+(l\nu+\lambda n)F+\lambda\nu G,$$

and $\qquad \mathbf{r}_{12}^2=M^2+m^2E+2m\mu F+\mu^2 G.$

It is also easily verified that

$$\mathbf{r}_{12}^2-\mathbf{r}_{11}\cdot\mathbf{r}_{22}=\tfrac{1}{2}\,(E_{22}+G_{11}-2F_{12}).$$

Adding the first and third, and subtracting the second, we obtain the required formula, which may be written

$$LN-M^2=\tfrac{1}{2}\,(2F_{12}-E_{22}-G_{11})+(m^2E+2m\mu F+\mu^2 G)$$
$$-\{lnE+(l\nu+\lambda n)F+\lambda\nu G\}\dots\dots(5).$$

This is the *Gauss characteristic equation*. It is sometimes expressed in the alternative form

$$LN-M^2=\tfrac{1}{2}H\frac{\partial}{\partial u}\left\{\frac{F}{EH}\frac{\partial E}{\partial v}-\frac{1}{H}\frac{\partial G}{\partial u}\right\}$$
$$+\tfrac{1}{2}H\frac{\partial}{\partial v}\left\{\frac{2}{H}\frac{\partial F}{\partial u}-\frac{1}{H}\frac{\partial E}{\partial v}-\frac{F}{EH}\frac{\partial E}{\partial u}\right\}\dots\dots(6).$$

The equation shows that *the specific curvature K, which is equal to $(LN-M^2)/H^2$, is expressible in terms of the fundamental magni-*

tudes E, F, G and their derivatives of the first two orders. In this respect it differs from the first curvature.

Cor. Surfaces which have the same first order magnitudes *E, F, G* (irrespective of the second order magnitudes *L, M, N*) have the same specific curvature.

Ex. Verify the Gauss equation (5) for the surfaces in Examples 2, 3, 6, 7 of Art. 41.

43. Mainardi-Codazzi relations. In addition to the Gauss characteristic equation, there are two other independent relations between the fundamental magnitudes and their derivatives. These may be established as follows. If in the identity

$$\frac{\partial}{\partial v}\mathbf{r}_{11} = \frac{\partial}{\partial u}\mathbf{r}_{12}$$

we substitute the values of $\mathbf{r}_{11}$ and $\mathbf{r}_{12}$ given in (1), we obtain

$$L_2\mathbf{n} + l_2\mathbf{r}_1 + \lambda_2\mathbf{r}_2 + L\mathbf{n}_2 + l\mathbf{r}_{12} + \lambda\mathbf{r}_{22}$$
$$= M_1\mathbf{n} + m_1\mathbf{r}_1 + \mu_1\mathbf{r}_2 + M\mathbf{n}_1 + m\mathbf{r}_{11} + \mu\mathbf{r}_{12}.$$

If in this we substitute again from (1) the values of the second derivatives of **r**, and also for $\mathbf{n}_1$ and $\mathbf{n}_2$ from Art. 27, we obtain a vector identity, expressed in terms of the non-coplanar vectors $\mathbf{n}, \mathbf{r}_1, \mathbf{r}_2$. We may then equate coefficients of like vectors on the two sides, and obtain three scalar equations. By equating coefficients of **n**, for example, we have

$$L_2 + lM + \lambda N = M_1 + mL + \mu M,$$

that is $L_2 - M_1 = mL - (l - \mu)M - \lambda N \dots\dots\dots\dots(7).$

Similarly from the identity $\frac{\partial}{\partial v}\mathbf{r}_{12} = \frac{\partial}{\partial u}\mathbf{r}_{22}$, on substituting from (1) the values of $\mathbf{r}_{12}$ and $\mathbf{r}_{22}$ we obtain the relation

$$M_2\mathbf{n} + m_2\mathbf{r}_1 + \mu_2\mathbf{r}_2 + M\mathbf{n}_2 + m\mathbf{r}_{12} + \mu\mathbf{r}_{22}$$
$$= N_1\mathbf{n} + n_1\mathbf{r}_1 + \nu_1\mathbf{r}_2 + N\mathbf{n}_1 + n\mathbf{r}_{11} + \nu\mathbf{r}_{12}.$$

Substituting again for the second derivatives of **r** and for $\mathbf{n}_1, \mathbf{n}_2$ in terms of $\mathbf{n}, \mathbf{r}_1, \mathbf{r}_2$, and equating coefficients of **n** on the two sides of the identity, we obtain

$$M_2 + mM + \mu N = N_1 + nL + \nu M,$$

that is $M_2 - N_1 = nL - (m - \nu)M - \mu N \dots\dots\dots\dots(8).$

The formulae (7) and (8) are frequently called the *Codazzi equations*. But as Mainardi gave similar results twelve years earlier than Codazzi, they are more justly termed the *Mainardi-Codazzi relations*. Four other formulae are obtained by equating coefficients of $\mathbf{r}_1$ and of $\mathbf{r}_2$ in the two identities: but they are not independent. They are all deducible from (7) and (8) with the aid of the Gauss characteristic equation.

44. Alternative expression. The above relations may be expressed in a different form, which is sometimes more useful. By differentiating the relation $H^2 = EG - F^2$ with respect to the parameters, it is easy to verify that

$$H_1 = H(l + \mu),$$

and

$$H_2 = H(m + \nu).$$

Therefore

$$\frac{\partial}{\partial u}\left(\frac{N}{H}\right) = \frac{N_1}{H} - \frac{N}{H^2}H_1$$

$$= \frac{N_1}{H} - \frac{N}{H}(l + \mu),$$

and similarly

$$\frac{\partial}{\partial v}\left(\frac{M}{H}\right) = \frac{M_2}{H} - \frac{M}{H^2}H_2$$

$$= \frac{M_2}{H} - \frac{M}{H}(m + \nu).$$

Consequently

$$\frac{\partial}{\partial v}\left(\frac{M}{H}\right) - \frac{\partial}{\partial u}\left(\frac{N}{H}\right) = \frac{1}{H}(M_2 - N_1) - \frac{M}{H}(m + \nu) + \frac{N}{H}(l + \mu)$$

$$= (nL - 2mM + lN)/H \quad\ldots\ldots\ldots\ldots\ldots(9),$$

in virtue of (8). Similarly it may be proved that

$$\frac{\partial}{\partial u}\left(\frac{M}{H}\right) - \frac{\partial}{\partial v}\left(\frac{L}{H}\right) = (\nu L - 2\mu M + \lambda N)/H \quad\ldots\ldots(10).$$

The equations (9) and (10) are an alternative form of the Mainardi-Codazzi relations.

We have seen that if six functions E, F, G, L, M, N constitute the fundamental magnitudes of a surface, they are connected by the three differential equations called the Gauss characteristic equation and the Mainardi-Codazzi relations. Conversely Bonnet has proved the theorem: *When six fundamental magnitudes are given, satisfying the Gauss characteristic equation and the Mainardi-*

Codazzi relations, they determine a surface uniquely, except as to position and orientation in space†. The proof of the theorem is beyond the scope of this book, and we shall not have occasion to use it.

*45. Derivatives of the angle ω. The coefficients occurring in Gauss's formulae of Art. 41 may be used to express the derivatives of the angle ω between the parametric curves. On differentiating the relation

$$\tan \omega = \frac{H}{F}$$

with respect to u, we have

$$\sec^2 \omega \, \omega_1 = \frac{FH_1 - HF_1}{F^2}.$$

Then on substituting the value $\sec^2 \omega = EG/F^2$, and multiplying both sides by $2HF^2$, we find

$$2EGH\omega_1 = F(2HH_1) - 2F_1 H^2$$
$$= F \frac{\partial}{\partial u}(EG - F^2) - 2F_1(EG - F^2)$$
$$= F(E_1 G + EG_1) - 2F_1 EG$$
$$= -2H^2(\lambda G + mE).$$

Hence the formula $$\omega_1 = -H\left(\frac{\lambda}{E} + \frac{m}{G}\right) \dots\dots\dots(11).$$

And in a similar manner it may be shown that

$$\omega_2 = -H\left(\frac{\mu}{E} + \frac{n}{G}\right) \dots\dots\dots(12).$$

EXAMPLES VII

1. Show that the other four relations, similar to the Mainardi-Codazzi relations, obtainable by equating coefficients of r_1 and of r_2 in the proof of Art. 43, are equivalent to

$$FK = m_1 - l_2 + m\mu - n\lambda,$$
$$FK = \mu_2 - \nu_1 + m\mu - n\lambda,$$
$$EK = \lambda_2 - \mu_1 + l\mu - m\lambda + \lambda\nu - \mu^2,$$
$$GK = n_1 - m_2 + ln - m^2 + m\nu - n\mu.$$

2. Prove that these formulae may be deduced from the Gauss characteristic equation and the Mainardi-Codazzi relations.

† Forsyth, *Differential Geometry*, p. 50.

3. Prove the relations

$$\frac{\partial}{\partial v}\left(\frac{H\lambda}{E}\right) - \frac{\partial}{\partial u}\left(\frac{H\mu}{E}\right) = HK,$$

$$\frac{\partial}{\partial u}\left(\frac{Hn}{G}\right) - \frac{\partial}{\partial v}\left(\frac{Hm}{G}\right) = HK,$$

using the formulae in Ex. 1.

4. If ω is the angle between the parametric curves, prove that

$$-\omega_{12} = \frac{\partial}{\partial u}\left(\frac{H\mu}{E}\right) + \frac{\partial}{\partial v}\left(\frac{Hm}{G}\right) + HK$$

$$= \frac{\partial}{\partial u}\left(\frac{Hn}{G}\right) + \frac{\partial}{\partial v}\left(\frac{H\lambda}{E}\right) - HK.$$

5. If the asymptotic lines are taken as parametric curves, show that the Mainardi-Codazzi relations become

$$\frac{M_1}{M} = l - \mu, \qquad \frac{M_2}{M} = \nu - m.$$

Hence deduce that (cf. Art. 44)

$$2l = \frac{H_1}{H} + \frac{M_1}{M}, \qquad 2\mu = \frac{H_1}{H} - \frac{M_1}{M},$$

$$2m = \frac{H_2}{H} - \frac{M_2}{M}, \qquad 2\nu = \frac{H_2}{H} + \frac{M_2}{M}.$$

6. When the parametric curves are null lines, show that the Mainardi-Codazzi relations may be expressed

$$\frac{L_2}{M} = \frac{\partial}{\partial u}\log\frac{M}{F}, \qquad \frac{N_1}{M} = \frac{\partial}{\partial v}\log\frac{M}{F},$$

and the Gauss characteristic equation as

$$LN - M^2 = F_{12} - \frac{F_1 F_2}{F}.$$

7. When the linear element is of the form

$$ds^2 = \phi\,(du^2 + dv^2),$$

the Mainardi-Codazzi relations are

$$L_2 - M_1 = \frac{1}{2}\frac{\phi_2}{\phi}(L + N),$$

$$N_1 - M_2 = \frac{1}{2}\frac{\phi_1}{\phi}(L + N),$$

and the Gauss equation

$$LN - M^2 = \frac{1}{2\phi}(\phi_1{}^2 + \phi_2{}^2) - \tfrac{1}{2}(\phi_{11} + \phi_{22}).$$

8. When the parametric curves are lines of curvature, deduce from equations (7) and (8) that

$$L_2 = Lm - N\lambda = \frac{1}{2}\left(\frac{L}{E} + \frac{N}{G}\right)E_2,$$

$$N_1 = N\mu - Ln = \frac{1}{2}\left(\frac{L}{E} + \frac{N}{G}\right)G_1.$$

9. Prove that, for any direction on a surface,

$$\mathbf{r}'' = \kappa_n \mathbf{n} + D_1 \mathbf{r}_1 + D_2 \mathbf{r}_2,$$

where

$$D_1 = l u'^2 + 2m u' v' + n v'^2 + u'',$$
$$D_2 = \lambda u'^2 + 2\mu u' v' + \nu v'^2 + v''.$$

10. With the notation of Ex. 9, show that the curvature of an asymptotic line, as given by (26) of Art. 38, may be expressed

$$H (D_2 u' - D_1 v').$$

Deduce the values $H\lambda/E^{\frac{3}{2}}$ and $- Hn/G^{\frac{3}{2}}$ for the curvatures of the parametric curves when these are asymptotic lines.

11. Prove the relations

$$E_1 = 2\,(lE + \lambda F), \quad F = mE + (l + \mu)\,F + \lambda G,$$
$$E_2 = 2\,(mE + \mu F), \quad F_2 = nE + (m + \nu)\,F + \mu G.$$

12. From the Gauss characteristic equation deduce that, when the parametric curves are orthogonal,

$$K = - \frac{1}{\sqrt{EG}} \left[\frac{\partial}{\partial u} \left(\frac{1}{\sqrt{E}} \frac{\partial \sqrt{G}}{\partial u} \right) + \frac{\partial}{\partial v} \left(\frac{1}{\sqrt{G}} \frac{\partial \sqrt{E}}{\partial v} \right) \right].$$

This formula is important

CHAPTER VI

GEODESICS AND GEODESIC PARALLELS

GEODESICS

46. Geodesic property. A *geodesic line*, or briefly a *geodesic*, on a surface may be defined as a curve whose osculating plane at each point contains the normal to the surface at that point. It follows that *the principal normal to the geodesic coincides with the normal to the surface;* and we agree to take it also in the same sense. The curvature of a geodesic is therefore the normal curvature of the surface in the direction of the curve, and has the value

$$\kappa = Lu'^2 + 2Mu'v' + Nv'^2 \quad \dots\dots\dots\dots\dots(1),$$

by Art. 28, the dashes denoting derivatives with respect to the arc-length s of the curve.

Moreover, of all plane sections through a given tangent line to the surface, the normal section has the least curvature, by Meunier's theorem. Therefore of all sections through two consecutive points P, Q on the surface, the normal section makes the length of the arc PQ a minimum. But this is the arc of the geodesic through P, Q. Hence a geodesic is sometimes defined as *the path of shortest distance* on the surface between two given points on it. Starting with this definition we may reverse the argument, and deduce the property that the principal normal to the geodesic coincides with the normal to the surface. The same may be done by the Calculus of Variations, or by statical considerations in the following manner. The path of shortest distance between two given points on the surface is the curve along which a flexible string would lie, on the (smooth) convex side of the surface, tightly stretched between the two points. Now the only forces on an element of the string are the tensions at its extremities and the reaction normal to the surface. But the tensions are in the osculating plane of the element, and therefore so also is the reaction by the condition of equilibrium. Thus the normal to the surface coincides with the principal normal to the curve.

47. Equations of geodesics. From the defining property of geodesics, and the Serret-Frenet formulae, it follows that

$$\mathbf{r}'' = \kappa \mathbf{n} \qquad\qquad\qquad\dots\dots\dots(2),$$

which may be expanded, as in Art. 28,

$$\mathbf{r}_1 u'' + \mathbf{r}_2 v'' + \mathbf{r}_{11} u'^2 + 2\mathbf{r}_{12} u' v' + \mathbf{r}_{22} v'^2 = \kappa \mathbf{n}.$$

Forming the scalar product of each side with $\mathbf{r}_1$ and $\mathbf{r}_2$ successively, we have

$$\left.\begin{array}{l} Eu'' + Fv'' + \tfrac{1}{2} E_1 u'^2 + E_2 u' v' + (F_2 - \tfrac{1}{2} G_1) v'^2 = 0 \\ Fu'' + Gv'' + (F_1 - \tfrac{1}{2} E_2) u'^2 + G_1 u' v' + \tfrac{1}{2} G_2 v'^2 = 0 \end{array}\right\} \quad\dots(3).$$

These are the *general differential equations of geodesics* on a surface. They are clearly equivalent to the equations

$$\left.\begin{array}{l} \dfrac{d}{ds}(Eu' + Fv') = \tfrac{1}{2}(E_1 u'^2 + 2F_1 u' v' + G_1 v'^2) \\[2mm] \dfrac{d}{ds}(Fu' + Gv') = \tfrac{1}{2}(E_2 u'^2 + 2F_2 u' v' + G_2 v'^2) \end{array}\right\} \quad\dots\dots(4).$$

A third form, which is sometimes more convenient, may be found by solving (3) for u'' and v'', thus obtaining

$$\left.\begin{array}{l} u'' + lu'^2 + 2mu'v' + nv'^2 = 0 \\ v'' + \lambda u'^2 + 2\mu u'v' + \nu v'^2 = 0 \end{array}\right\} \quad\dots\dots\dots(5),$$

where l, λ etc. are the coefficients of Art. 41.

A curve on the surface is, however, determined by a single relation between the parameters. Hence the above pair of differential equations may be replaced by a single relation between u, v. If, for example, we take the equations (5), multiply the first by $\dfrac{dv}{du}\left(\dfrac{ds}{du}\right)^2$, the second by $\left(\dfrac{ds}{du}\right)^2$ and subtract, we obtain the *single differential equation of geodesics* in the form

$$\frac{d^2 v}{du^2} = n \left(\frac{dv}{du}\right)^3 + (2m - \nu)\left(\frac{dv}{du}\right)^2 + (l - 2\mu)\frac{dv}{du} - \lambda \quad\dots(6).$$

Now from the theory of differential equations it follows that there exists a unique integral v of this equation which takes a given value v_0 when $u = u_0$, and whose derivative dv/du also takes a given value when $u = u_0$. Thus *through each point of the surface there passes a single geodesic in each direction.* Unlike lines of curvature and asymptotic lines, geodesics are not determined uniquely or in pairs at a point by the nature of the surface. Through any point pass an infinite number of geodesics, each geodesic being determined by its direction at the point.

The equations of geodesics involve only the magnitudes of the first order, E, F, G, and their derivatives. Hence if the surface is deformed without stretching or tearing, so that the length ds of each arc element is unaltered, the *geodesics remain geodesics on the deformed surface*. In particular, when a developable surface is developed into a plane, the geodesics on the surface become straight lines on the plane. This agrees with the fact that a straight line is the path of shortest distance between two given points on the plane.

From (6) it follows immediately that the parametric curves $v =$ const. will be geodesics if $\lambda = 0$. Similarly the curves $u =$ const. will be geodesics if $n = 0$. Hence, *if the parametric curves are orthogonal* $(F = 0)$, *the curves* $v =$ const. *will be geodesics provided E is a function of u only, and the curves* $u =$ const. *will be geodesics if G is a function of v only.*

Ex. 1. On the right helicoid given by

$$x = u \cos \phi, \quad y = u \sin \phi, \quad z = c\phi,$$

we have seen (Ex. 1, Art. 26) that

$$E = 1, \quad F = 0, \quad G = u^2 + c^2, \quad H^2 = u^2 + c^2$$

Therefore the coefficients of Art. 41 have the values

$$l = 0, \quad m = 0, \qquad n = -u,$$
$$\lambda = 0, \quad \mu = u/(u^2 + c^2), \quad \nu = 0.$$

The equations (5) for the geodesics become

$$\left. \begin{array}{r} u'' - u\phi'^2 = 0 \\ (u^2 + c^2) \, \phi'' + 2u \, u'\phi' = 0 \end{array} \right\}$$

From the second of these it follows that

$$(u^2 + c^2) \frac{d\phi}{ds} = \text{const.} = h \text{ (say).}$$

But for any arc on the surface

$$ds^2 = du^2 + (u^2 + c^2) \, d\phi^2.$$

Hence, for the arc of a geodesic,

$$(u^2 + c^2)^2 \, d\phi^2 = h^2 du^2 + (u^2 + c^2) \, h^2 d\phi^2,$$

and therefore

$$\frac{du}{d\phi} = \pm \frac{1}{h} \sqrt{(u^2 + c^2)(u^2 + c^2 - h^2)}.$$

This is a first integral of the differential equation of geodesics. The complete integral may be found in terms of elliptic functions.

Ex. 2. When the equation of the surface is given in Monge's form $z = f(x, y)$, we have seen (Art. 33) that, with x, y as parameters,

$$E = 1 + p^2, \quad F = pq, \quad G = 1 + q^2, \quad H^2 = 1 + p^2 + q^2.$$

Therefore, by Art. 41, $\quad l = \dfrac{pr}{H^2}, \quad m = \dfrac{ps}{H^2}, \quad n = \dfrac{pt}{H^2},$

$$\lambda = \frac{qr}{H^2}, \quad \mu = \frac{qs}{H^2}, \quad \nu = \frac{qt}{H^2}.$$

The equation (6) for geodesics then takes the form

$$(1 + p^2 + q^2)\frac{d^2 y}{dx^2} = pt\left(\frac{dy}{dx}\right)^3 + (2ps - qt)\left(\frac{dy}{dx}\right)^2 + (pr - 2qs)\frac{dy}{dx} - rq$$

$$= \left(p\frac{dy}{dx} - q\right)\left\{ t\left(\frac{dy}{dx}\right)^2 + 2s\frac{dy}{dx} + r \right\}.$$

48. Surface of revolution. On the surface of revolution

$$x = u\cos\phi, \quad y = u\sin\phi, \quad z = f(u),$$

we have seen (Art. 34) that with u, ϕ as parameters

$$E = 1 + f_1^2, \quad F = 0, \quad G = u^2, \quad H^2 = u^2(1 + f_1^2).$$

Therefore, by Art. 41,

$$\lambda = 0, \quad \mu = \frac{1}{u}, \quad \nu = 0.$$

The second of equations (5) for geodesics then takes the form

$$\frac{d^2\phi}{ds^2} + \frac{2}{u}\frac{du}{ds}\frac{d\phi}{ds} = 0.$$

On multiplication by u^2 this equation becomes exact, and has for its integral

$$u^2 \frac{d\phi}{ds} = h \quad\quad\quad\quad\quad\ldots\ldots\ldots(7),$$

where h is a constant. Or, if ψ is the angle at which the geodesic cuts the meridian, we may write this result

$$u\sin\psi = h \ldots\ldots\ldots\ldots\ldots(7'),$$

a theorem due to Clairaut. This is a first integral of the equation of geodesics, involving one arbitrary constant h.

To obtain the complete integral we observe that, for any arc on the surface,

$$ds^2 = (1 + f_1^2)\,du^2 + u^2 d\phi^2,$$

and therefore, by (7), for the arc of a geodesic,

$$u^4 d\phi^2 = h^2(1 + f_1^2)\,du^2 + h^2 u^2 d\phi^2,$$

so that

$$d\phi = \pm \frac{h}{u}\sqrt{\frac{1 + f_1^2}{u^2 - h^2}}\,du.$$

Thus $$\phi = C \pm h \int \frac{1}{u} \sqrt{\frac{1+f_1^2}{u^2 - h^2}}\, du \dots\dots\dots\dots(8),$$

involving the two arbitrary constants C and h, is the complete integral of the equation of geodesics on a surface of revolution.

Cor. It follows from (7') that h is the minimum distance from the axis of a point on the geodesic, and is attained where the geodesic cuts a meridian at right angles.

Ex. 1. The geodesics on a circular cylinder are helices.

For from (7'), since u is constant ψ is constant. Thus the geodesics cut the generators at a constant angle, and are therefore helices.

Ex. 2. In the case of a right circular cone of semi-vertical angle a, show that the equation (8) for geodesics is equivalent to

$$u = h \sec (\phi \sin a + \beta),$$

where h and β are constants.

Ex. 3. The perpendicular from the vertex of a right circular cone, to a tangent to a given geodesic, is of constant length.

49. Torsion of a geodesic. If $\mathbf{r}$ is a point on the geodesic, $\mathbf{r}'$ is the unit tangent and the principal normal is the unit normal $\mathbf{n}$ to the surface. Hence the unit binormal is

$$\mathbf{b} = \mathbf{r}' \times \mathbf{n}.$$

Differentiation with respect to the arc-length gives for the torsion of the geodesic

$$-\tau \mathbf{n} = \mathbf{r}'' \times \mathbf{n} + \mathbf{r}' \times \mathbf{n}'.$$

The first term in the second member is zero because $\mathbf{r}''$ is parallel to $\mathbf{n}$. Hence

$$\tau \mathbf{n} = \mathbf{n}' \times \mathbf{r}' \dots\dots\dots\dots\dots\dots(9),$$

and therefore $$\tau = [\mathbf{n}, \mathbf{n}', \mathbf{r}'] \dots\dots\dots\dots\dots(10).$$

This expression for the torsion of a geodesic is identical with that found in Art. 38 for the torsion of an asymptotic line. The geodesic which touches a curve at any point is often called its *geodesic tangent* at that point. Hence *the torsion of an asymptotic line is equal to the torsion of its geodesic tangent.*

Further, the expression $[\mathbf{n}, \mathbf{n}', \mathbf{r}']$ vanishes for a principal direction (Art. 29). Hence *the torsion of a geodesic vanishes where it touches a line of curvature.* It also follows from (10) that *if a geodesic is a plane curve it is a line of curvature; and, conversely, if a geodesic is a line of curvature it is also a plane curve.*

The triple product $[\mathbf{n}, \mathbf{n}', \mathbf{r}']$ may be expanded, as in Art. 29, by writing $\mathbf{n}' = \mathbf{n}_1 u' + \mathbf{n}_2 v'$ and $\mathbf{r}' = \mathbf{r}_1 u' + \mathbf{r}_2 v'$. The formula for the torsion of a geodesic then becomes

$$\tau = \frac{1}{H} \{(EM - FL) u'^2 + (EN - GL) u'v' + (FN - GM) v'^2\} \quad (11).$$

This may be expressed in terms of the inclination of the geodesic to the principal directions. Let the lines of curvature be taken as parametric curves. Then

$$F = M = 0, \quad H^2 = EG,$$

and the last formula becomes

$$\tau = \sqrt{EG}\, u'v' \left(\frac{N}{G} - \frac{L}{E}\right).$$

But (Art. 24 and Note I) if ψ is the inclination of the geodesic to the line of curvature $v = $ constant,

$$\sqrt{E}\, u' = \cos \psi, \quad \sqrt{G}\, v' = \sin \psi.$$

Also the principal curvatures are

$$\kappa_a = L/E, \quad \kappa_b = N/G.$$

Hence the formula for the torsion of the geodesic becomes

$$\tau = \cos \psi \sin \psi \, (\kappa_b - \kappa_a) \ldots\ldots\ldots\ldots\ldots(12).$$

From this it follows that *two geodesics at right angles have their torsions equal in magnitude but opposite in sign*. Further, besides vanishing in the principal directions, *the torsion of a geodesic vanishes at an umbilic*. And, of all geodesics through a given point, those which bisect the angles between the lines of curvature have the greatest torsion.

The *curvature* of a geodesic is the normal curvature in its direction. Its value, as given by Euler's theorem (Art. 31), is therefore

$$\kappa = \kappa_a \cos^2 \psi + \kappa_b \sin^2 \psi \ldots\ldots\ldots\ldots\ldots(13).$$

Ex. 1. If κ, τ are the curvature and torsion of a geodesic, prove that

$$\tau^2 = (\kappa - \kappa_a)(\kappa_b - \kappa).$$

Also, if the surface is developable ($\kappa_a = 0$), show that

$$\kappa = \tau \tan \psi.$$

Ex. 2. Deduce from (12) that the torsions of the two asymptotic lines at a point are equal in magnitude and opposite in sign.

Ex. 3. Prove that the torsion of a geodesic is equal to

$$\frac{1}{H}\begin{vmatrix} Eu' + Fv' & Fu' + Gv' \\ Lu' + Mv' & Mu' + Nv' \end{vmatrix}.$$

Ex. 4. Prove that, with the notation of Art. 49 for a geodesic,

$$\kappa \cos \psi - \tau \sin \psi = \kappa_a \cos \psi,$$

$$\kappa \sin \psi + \tau \cos \psi = \kappa_b \sin \psi.$$

CURVES IN RELATION TO GEODESICS

50. Bonnet's theorem. Let C be any curve drawn on the surface, $\mathbf{r}'$ its unit tangent, $\bar{\mathbf{n}}$ its principal normal, τ its torsion, and W the torsion of the geodesic which touches it at the point considered. We define the *normal angle* ϖ of the curve as the angle from $\bar{\mathbf{n}}$ to the normal $\mathbf{n}$ to the surface, in the positive sense

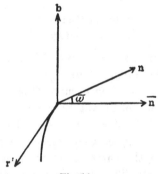

Fig. 14.

for a rotation about $\mathbf{r}'$. Thus ϖ is positive if the rotation from $\bar{\mathbf{n}}$ to $\mathbf{n}$ is in the sense from $\bar{\mathbf{n}}$ to the binormal $\mathbf{b}$; negative if in the opposite sense. Then at any point of the curve these quantities are connected by the relation

$$\frac{d\varpi}{ds} + \tau = W \quad \dots\dots\dots\dots\dots\dots(14).$$

This may be proved in the following manner. By (9) of the previous Art. we have $W\mathbf{n} = \mathbf{n}' \times \mathbf{r}'$. The unit binormal to the curve is $\mathbf{b} = \mathbf{r}' \times \bar{\mathbf{n}}$, and

$$\cos \varpi = \bar{\mathbf{n}} \cdot \mathbf{n}, \quad \sin \varpi = \mathbf{b} \cdot \mathbf{n}.$$

Differentiating this last equation, we have

$$\cos \varpi \, \frac{d\varpi}{ds} = \mathbf{b}' \cdot \mathbf{n} + \mathbf{b} \cdot \mathbf{n}'$$

$$= -\tau \bar{\mathbf{n}} \cdot \mathbf{n} + \mathbf{r}' \times \bar{\mathbf{n}} \cdot \mathbf{n}$$

$$= -\tau \cos \varpi + W \mathbf{n} \cdot \bar{\mathbf{n}}$$

$$= (-\tau + W) \cos \varpi.$$

Hence the formula $\dfrac{d\varpi}{ds} + \tau = W,$

expressing a result due to Bonnet. Since W is the torsion of the geodesic tangent, it follows that *the quantity* $\left(\dfrac{d\varpi}{ds} + \tau\right)$ *has the same value for all curves touching at the point considered.* The formula also shows that ϖ' is the torsion of the geodesic tangent relative to the curve C; or that $-\varpi'$ is that of C relative to the geodesic tangent.

Ex. Prove (14) by differentiating the formula

$$\cos \varpi = \bar{\mathbf{n}} \cdot \mathbf{n}.$$

51. Joachimsthal's theorems. We have seen that the torsion W of the geodesic tangent to a line of curvature vanishes at the point of contact. If then a curve C on the surface is both a plane curve and a line of curvature, $\tau = 0$ and $W = 0$; and therefore, in virtue of (14), $\varpi' = 0$. Consequently its plane cuts the surface at a constant angle. Conversely, if a plane cuts a surface at a constant angle, the curve of intersection has zero torsion, so that $\tau = 0$ and $\varpi' = 0$. Therefore, in virtue of (14), W vanishes identically, showing that the curve is a line of curvature. Similarly if ϖ is constant and the curve is a line of curvature, τ must vanish, and the curve is plane. Hence *if a curve on a surface has two of the following properties it also has the third: (a) it is a line of curvature, (b) it is a plane curve, (c) its normal angle is constant.*

Moreover, *if the curve of intersection of two surfaces is a line of curvature on each, the surfaces cut at a constant angle.* Let ϖ and ϖ_0 be the normal angles of the curve for the two surfaces. Then since the torsion W of the geodesic tangent vanishes on both surfaces,

$$\frac{d\varpi}{ds} + \tau = 0, \quad \frac{d\varpi_0}{ds} + \tau = 0.$$

Hence $$\frac{d}{ds}(\varpi - \varpi_0) = 0,$$

so that $$\varpi - \varpi_0 = \text{const.}$$

Thus the surfaces cut at a constant angle.

Similarly, *if two surfaces cut at a constant angle, and the curve of intersection is a line of curvature on one, it is a line of curvature on the other also.* For since

$$\varpi - \varpi_0 = \text{const.}$$

it follows that $$\frac{d\varpi}{ds} = \frac{d\varpi_0}{ds}.$$

Hence, by (14), if W and W_0 are the torsions of the geodesic tangents on the two surfaces,

$$W - \tau = W_0 - \tau,$$

so that $$W = W_0.$$

If then W vanishes, so does W_0, showing that the curve is a line of curvature on the second surface also. The above theorems are due to Joachimsthal. The last two were proved in Art. 29 by another method.

Further, we can prove theorems for *spherical lines of curvature,* similar to those proved above for plane lines of curvature. Geodesics on a sphere are great circles, and therefore plane curves. Their torsion W_0 therefore vanishes identically. Hence for any curve on a sphere, if ϖ_0 is its normal angle,

$$\frac{d\varpi_0}{ds} + \tau = 0.$$

Suppose then that a surface is cut by a sphere in a line of curvature. Then since the torsion W of the geodesic tangent to a line of curvature is zero, we have on this surface also

$$\frac{d\varpi}{ds} + \tau = 0.$$

From these two equations it follows that

$$\frac{d}{ds}(\varpi - \varpi_0) = 0,$$

and therefore $$\varpi - \varpi_0 = \text{const.}$$

Hence *if the curve of intersection of a sphere and another surface is a line of curvature on the latter, the two surfaces cut at a constant angle.*

Conversely, *if a sphere cuts a surface at a constant angle, the curve of intersection is a line of curvature on the surface.* For

$$\frac{d\varpi}{ds} = \frac{d\varpi_0}{ds},$$

and therefore $\qquad \tau = \tau - W.$

Thus W vanishes identically, and the curve is a line of curvature.

52. Vector curvature. The curvature of a curve, as defined in Art. 2, is a scalar quantity equal to the arc-rate of turning of the tangent. This is the magnitude of the *vector curvature*, which may be defined as *the arc-rate of change of the unit tangent*. It is therefore equal to $\mathbf{t}'$ or $\kappa\mathbf{n}$. Thus the direction of the vector curvature is parallel to the principal normal. The scalar curvature κ is the measure of the vector curvature, the positive direction along the principal normal being that of the unit vector $\mathbf{n}$.

If two curves, C and C_0, touch each other at P, we may define their *relative curvature* at this point as the difference of their vector curvatures. Let $\mathbf{t}$ be their common unit tangent at P, and

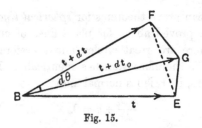

Fig. 15.

$\mathbf{t} + d\mathbf{t}, \mathbf{t} + d\mathbf{t}_0$ the unit tangents at consecutive points distant ds along the curves from P. If BE, BF, BG represent these unit vectors, the vector GF is equal to $d\mathbf{t} - d\mathbf{t}_0$. The (vector) curvature of C relative to C_0 is then

$$\frac{d\mathbf{t}}{ds} - \frac{d\mathbf{t}_0}{ds} = \frac{d\mathbf{t} - d\mathbf{t}_0}{ds} = \frac{GF}{ds}.$$

If $d\theta$ is the angle GBF, the magnitude of the relative curvature is $d\theta/ds$, the arc-rate of deviation of their tangents.

53. Geodesic curvature. Consider any curve C drawn on a surface. We define the *geodesic curvature* of the curve at a point P as its curvature relative to the geodesic which touches it at P.

Now the vector curvature of the curve is $\mathbf{r}''$, and the resolved part of this in the direction of the normal to the surface is $\mathbf{n} \cdot \mathbf{r}''$, or κ_n by Meunier's theorem. But the vector curvature of the geodesic is normal to the surface, and its magnitude is also κ_n. That is to say, the curvature of the geodesic is the normal resolved part of the vector curvature of C. Hence the curvature of C relative to the geodesic is its resolved part tangential to the surface. This tangential resolute is sometimes called the *tangential curvature* of C, but more frequently its *geodesic curvature*. As a vector it is given by

$$\mathbf{r}'' - \mathbf{n} \cdot \mathbf{r}'' \mathbf{n} \quad \text{or} \quad \mathbf{r}'' - \kappa_n \mathbf{n} \quad \dots\dots\dots\dots(15).$$

Its magnitude must be regarded as positive when the deviation of C from the geodesic tangent is in the positive sense for a rotation about the normal to the surface. Thus we must take the resolved part of the vector curvature $\mathbf{r}''$ in the direction of the unit vector $\mathbf{n} \times \mathbf{r}'$. Hence the magnitude of the geodesic curvature is $\mathbf{n} \times \mathbf{r}' \cdot \mathbf{r}''$. Denoting it by κ_g we have

$$\kappa_g = [\mathbf{n}, \mathbf{r}', \mathbf{r}''] \dots\dots\dots\dots\dots\dots(16).$$

A variation of this formula is obtained by writing $\mathbf{n} = \mathbf{r}_1 \times \mathbf{r}_2/H$. Then

$$[\mathbf{n}, \mathbf{r}', \mathbf{r}''] = \frac{1}{H}(\mathbf{r}_1 \times \mathbf{r}_2) \times \mathbf{r}' \cdot \mathbf{r}''$$

$$= \frac{1}{H}(\mathbf{r}_1 \cdot \mathbf{r}' \mathbf{r}_2 - \mathbf{r}_2 \cdot \mathbf{r}' \mathbf{r}_1) \cdot \mathbf{r}'',$$

so that

$$\kappa_g = \frac{1}{H}(\mathbf{r}_1 \cdot \mathbf{r}' \mathbf{r}_2 \cdot \mathbf{r}'' - \mathbf{r}_2 \cdot \mathbf{r}' \mathbf{r}_1 \cdot \mathbf{r}'') \dots\dots\dots(17).$$

It is also clear from the above argument that, if κ is the curvature of the curve C, and ϖ its normal angle,

$$\left. \begin{array}{l} \kappa_g = \kappa \sin \varpi \\ \kappa_n = \kappa \cos \varpi \end{array} \right\} \dots\dots\dots\dots\dots(18).$$

while

Hence

$$\left. \begin{array}{l} \kappa^2 = \kappa_g^2 + \kappa_n^2 \\ \kappa_g = \kappa_n \tan \varpi \end{array} \right\} \dots\dots\dots\dots(19).$$

and

All these expressions for κ_g vanish when C is a geodesic. For then $\mathbf{r}''$ is parallel to $\mathbf{n}$, and therefore perpendicular to $\mathbf{r}_1$ and $\mathbf{r}_2$, while ϖ is zero. This means simply that the curvature of a geodesic relative to itself is zero.

It will be noticed that the expression $[\mathbf{n}, \mathbf{r}', \mathbf{r}'']$ for the geodesic curvature is the same as that found in Art. 38 for the curvature of an asymptotic line. This is due to the fact that the osculating plane for an asymptotic line is the tangent plane to the surface, while the curvature of the geodesic tangent, being the normal curvature in the asymptotic direction, is zero. Thus *the curvature of an asymptotic line is equal to its geodesic curvature.*

54. Other formulae for κ_g. From (16) or (17) we may deduce an expansion for the geodesic curvature in terms of u', u'' etc. For instance, on substitution of the values of $\mathbf{r}'$ and $\mathbf{r}''$ in terms of these, (17) becomes

$$\kappa_g = \frac{1}{H}\left(Eu' + Fv'\right)\left\{Fu'' + Gv'' + (F_1 - \tfrac{1}{2}E_2)\,u'^2 + G_1 u'v' + \tfrac{1}{2}G_2 v'^2\right\}$$

$$- \frac{1}{H}\left(Fu' + Gv'\right)\left\{Eu'' + Fv'' + \tfrac{1}{2}E_1 u'^2 + E_2 u'v' + (F_2 - \tfrac{1}{2}G_1)\,v'^2\right\},$$

which may also be written

$$\kappa_g = Hu'\,(v'' + \lambda u'^2 + 2\mu u'v' + \nu v'^2)$$
$$- Hv'\,(u'' + lu'^2 + 2mu'v' + nv'^2)\dots(20),$$

each part of which vanishes for a geodesic, in virtue of (5).

In particular for the parametric curve $v = \text{const.}$ we have $v' = v'' = 0$, and the geodesic curvature κ_{gu} of this curve is therefore equal to $Hu'\lambda u'^2$, which may be written

$$\kappa_{gu} = H\lambda E^{-\frac{3}{2}}.$$

Similarly the geodesic curvature κ_{gv} of the curve $u = \text{const.}$ has the value

$$\kappa_{gv} = - HnG^{-\frac{3}{2}}.$$

When the parametric curves are *orthogonal*, these become

$$\kappa_{gu} = - \frac{E_2}{2E\sqrt{G}}, \quad \kappa_{gv} = \frac{G_1}{2G\sqrt{E}}.$$

From these formulae we may deduce the results, already noticed in Art. 47, that the curves $v = \text{const.}$ will be geodesics provided $\lambda = 0$, and the curves $u = \text{const.}$ provided $n = 0$. When the parametric curves are orthogonal, these conditions are $E_2 = 0$ and $G_1 = 0$; so that the curves $v = \text{const.}$ will be geodesics if E is a function of u only; and the curves $u = \text{const.}$ if G is a function of v only.

Another formula for the geodesic curvature of a curve may be found in terms of the arc-rate of increase of its inclination to the parametric curves. Let θ be the inclination of the curve to the parametric curve $v = $ const., measured in the positive sense. Then since, by Art. 24 and Note I

$$Eu' + Fv' = \sqrt{E} \cos \theta.$$

we have on differentiation

$$\frac{d}{ds}(Eu' + Fv') = \frac{d\sqrt{\bar{E}}}{ds} \cos \theta - \sqrt{\bar{E}} \sin \theta \frac{d\theta}{ds}$$

$$= \frac{1}{2E}(E_1 u' + E_2 v')(Eu' + Fv') - Hv' \frac{d\theta}{ds}.$$

Now, if the curve is a geodesic, the first member of this equation is equal to

$$\tfrac{1}{2}(E_1 u'^2 + 2F_1 u' v' + G_1 v'^2).$$

On substitution of this value we find for a geodesic

$$H \frac{d\theta}{ds} = -\frac{H^2 \lambda}{E} u' - \frac{H^2 \mu}{E} v'.$$

Thus the rate of increase of the inclination of a geodesic to the parametric curve $v = $ const. is given by

$$\frac{d\theta}{ds} = -\frac{H}{E}(\lambda u' + \mu v').$$

Now the geodesic curvature of a curve C is tangential to the surface, and its magnitude is the arc-rate of deviation of C from its geodesic tangent. This is equal to the difference of the values of $d\theta/ds$ for the curve and for its geodesic tangent. But its value for the geodesic has just been found. Hence, if $d\theta/ds$ denotes its value for the curve C, the geodesic curvature of C is given by

$$\kappa_g = \frac{d\theta}{ds} + \frac{H}{E}(\lambda u' + \mu v') \dots\dots\dots\dots\dots(21).$$

Or, if ϑ is the inclination of the parametric curve $u = $ const. to the curve C (Fig. 11, Art. 24), we may write this

$$\kappa_g = \frac{d\theta}{ds} + \frac{\lambda \sqrt{G}}{E} \sin \vartheta + \frac{\mu}{\sqrt{E}} \sin \theta \dots\dots\dots\dots(22).$$

In the particular case when the parametric curves are orthogonal, $\sin \vartheta = \cos \theta$. Also the coefficient of $\sin \vartheta$ becomes equal to the geodesic curvature of the curve $v = $ const., and the coefficient of $\sin \theta$ to that of the curve $u = $ const. Denoting these by κ_{gu} and κ_{gv} respectively, we have *Liouville's formula*

$$\kappa_g = \frac{d\theta}{ds} + \kappa_{gu} \cos \theta + \kappa_{gv} \sin \theta \dots\dots\dots\dots\dots(23).$$

*55. Examples.

(1) *Bonnet's formula* for the geodesic curvature of the curve $\phi\,(u,\,v)=0$.
By differentiation we have $\qquad \phi_1 u' + \phi_2 v' = 0$ ....(a),

so that
$$\frac{u'}{\phi_2} = \frac{v'}{-\phi_1} = \frac{1}{\Theta},$$

where
$$\Theta = \sqrt{E\phi_2{}^2 - 2F\phi_1\phi_2 + G\phi_1{}^2}.$$

Again differentiating (a) we find
$$\phi_1 u'' + \phi_2 v'' + \phi_{11} u'^2 + 2\phi_{12} u'v' + \phi_{22} v'^2 = 0,$$
which may be written
$$\Theta\,(u'v'' - v'u'') + \phi_{11}u'^2 + 2\phi_{12}u'v' + \phi_{22}v'^2 = 0.$$
By means of these relations we find that
$$\frac{\partial}{\partial u}\left(\frac{F\phi_2 - G\phi_1}{\Theta}\right) + \frac{\partial}{\partial v}\left(\frac{F\phi_1 - E\phi_2}{\Theta}\right)$$
$$= H^2 u'\,(v'' + \lambda u'^2 + 2\mu u'v' + \nu v'^2) - H^2 v'\,(u'' + lu'^2 + 2m u'v' + n v'^2)$$
$$= H\kappa_g.$$
Hence Bonnet's formula for the geodesic curvature
$$\kappa_g = \frac{1}{H}\frac{\partial}{\partial u}\left(\frac{F\phi_2 - G\phi_1}{\Theta}\right) + \frac{1}{H}\frac{\partial}{\partial v}\left(\frac{F\phi_1 - E\phi_2}{\Theta}\right) \quad\ldots\ldots\ldots\ldots(\beta).$$

From this result we may deduce the geodesic curvature of a curve of the
family defined by the differential equation
$$Pdu + Qdv = 0 \quad\ldots\ldots\ldots\ldots\ldots\ldots\ldots\ldots(\gamma).$$
For, on comparing this equation with (a), we see that the required value is
$$\kappa_g = \frac{1}{H}\frac{\partial}{\partial u}\left(\frac{FQ - GP}{\sqrt{EQ^2 - 2FPQ + GP^2}}\right) + \frac{1}{H}\frac{\partial}{\partial v}\left(\frac{FP - EQ}{\sqrt{EQ^2 - 2FPQ + GP^2}}\right).$$

(2) Deduce the geodesic curvatures of the parametric curves from the results
of the previous exercise.

(3) A curve C touches the parametric curve $v = $ const. Find its curvature
relative to the parametric curve at the point of contact.

The relative curvature is the difference of their geodesic curvatures. The
geodesic curvature of C is got from (20) by putting $v' = 0$ and $u' = 1/\sqrt{E}$. Its
value is therefore $H\,(v'' + \lambda E^{-1})/\sqrt{E}$. But the geodesic curvature of $v = $ const.
is $H\lambda E^{-\frac{3}{2}}$. Hence the relative curvature is $Hv''/\sqrt{E}$.

(4) Find the geodesic curvature of the parametric lines on the surface
$$x = a\,(u+v), \quad y = b\,(u-v), \quad z = uv.$$

(5) Find the geodesic curvature of a parallel on a surface of revolution.

(6) Show that a twisted curve is a geodesic on its rectifying developable.
(The principal normal of the curve is normal to the surface.)

(7) Show that the evolutes of a twisted curve are geodesics on its polar
developable (Arts. 11 and 19).

(8) The radius of curvature of a geodesic on a cone of revolution varies as
the cube of the distance from the vertex.

(9) From the formula (16) deduce the geodesic curvature of the curve
$v = $ const., putting $\mathbf{r}' = \mathbf{r}_1/\sqrt{E}$.

GEODESIC PARALLELS

56. Geodesic parallels. Let a singly infinite family of geodesics on the surface be taken as parametric curves $v = \text{const.}$, and their orthogonal trajectories as the curves $u = \text{const.}$ Then $F = 0$, and the square of the linear element has the form

$$ds^2 = E\,du^2 + G\,dv^2.$$

Further, since the curves $v = \text{const.}$ are geodesics, E is a function of u alone (Art. 47). Hence, if we take $\int \sqrt{E}\,du$ as a new parameter u, we have

$$ds^2 = du^2 + G\,dv^2 \quad\dots\dots\dots\dots\dots\dots(24),$$

which is called the *geodesic form* for ds^2. Since E is now equal to unity, the length of an element of arc of a geodesic is du; and the length of a geodesic intercepted between the two trajectories $u = a$ and $u = b$ is

$$\int_a^b du = b - a.$$

This is the same for all geodesics of the family, and is called the *geodesic distance* between the two curves. On account of this property the orthogonal trajectories $u = \text{const.}$ are called *geodesic parallels*.

From the geodesic form for ds^2 we may easily deduce the property of minimum length characteristic of the arc of a geodesic joining two points on it. Consider, for example, the two points P, Q in which a geodesic is cut by the parallels $u = a, u = b$. The length of the arc of the geodesic joining the two points is $(b - a)$. For any other curve joining them the length of arc is

$$\int_P^Q ds = \int_P^Q \sqrt{du^2 + G\,dv^2} > \int_a^b du,$$

since G is positive. Thus the distance is least in the case of the geodesic.

With the above choice of parameters many results take a simpler form. Since G is positive it may be replaced by D^2, so that

$$ds^2 = du^2 + D^2\,dv^2 \quad\dots\dots\dots\dots\dots\dots(25).$$

Then since $F = 0$ and $E = 1$ we have $H^2 = G$, so that

$$l = 0, \quad m = 0, \qquad n = -\tfrac{1}{2}G_1 = -DD_1,$$

$$\lambda = 0, \quad \mu = \frac{1}{2}\frac{G_1}{G} = \frac{D_1}{D}, \quad \nu = \frac{1}{2}\frac{G_2}{G} = \frac{D_2}{D}.$$

The Gauss characteristic equation becomes

$$LN - M^2 = -\tfrac{1}{2}G_{11} + \frac{G_1^2}{4G} = -\sqrt{G}\,\frac{\partial^2 \sqrt{G}}{\partial u^2},$$

and therefore the specific curvature is

$$K = \frac{LN - M^2}{G} = -\frac{1}{\sqrt{G}}\frac{\partial^2 \sqrt{G}}{\partial u^2},$$

or $$K = -\frac{1}{D}\frac{\partial^2 D}{\partial u^2} \quad\dots\dots\dots\dots\dots\dots\dots(26).$$

The first curvature is $$J = L + \frac{N}{G}.$$

The general equations (4) of geodesics become

$$\left.\begin{aligned}u'' - DD_1 v'^2 &= 0\\ \frac{d}{ds}(D^2 v') - DD_2 v'^2 &= 0\end{aligned}\right\} \quad\dots\dots\dots\dots(27),$$

and the single equation (6) gives

$$D\frac{d^2 v}{du^2} + D^2 D_1 \left(\frac{dv}{du}\right)^3 + D_2 \left(\frac{dv}{du}\right)^2 + 2D_1 \frac{dv}{du} = 0.$$

Ex. *Beltrami's theorem.* Consider a singly infinite family of geodesics, cut by a curve C whose direction at any point P is conjugate to that of the geodesic through P. The tangents to the geodesics at the points of C generate a developable surface (Art. 35), and are tangents to its edge of regression. Beltrami's theorem is that *the centre of geodesic curvature at P, of that orthogonal trajectory of the geodesics which passes through this point, is the corresponding point on the edge of regression.*

Let the geodesics be taken as the curves $v=$const. and their orthogonal trajectories as the curves $u=$const. Then the square of the linear element has the geodesic form

$$ds^2 = du^2 + G\,dv^2.$$

The geodesic curvature of the parametric curve $u=$const. is, by Art. 54,

$$\kappa_g = \frac{1}{2G}\frac{\partial G}{\partial u}.$$

This is measured in the sense of the rotation from $\mathbf{r}_1$ to $\mathbf{r}_2$. Hence the distance ρ from P to the centre of geodesic curvature, measured in the direction $\mathbf{r}_1$, is given by

$$\frac{1}{\rho} = -\frac{1}{2G}\frac{\partial G}{\partial u}.$$

Let $\mathbf{r}$ be the position vector of the point P on the curve C, $\mathbf{R}$ that of the corresponding point Q on the edge of regression, and r the distance PQ, also measured in the direction $\mathbf{r}_1$. Then, since $E=1$,

$$\mathbf{R} = \mathbf{r} + r\mathbf{r}_1.$$

Along C the quantities are functions of the arc-length s of the curve. Hence, on differentiation,

$$\mathbf{R}' = (\mathbf{r}_1 u' + \mathbf{r}_2 v') + r' \mathbf{r}_1 + r(\mathbf{r}_{11} u' + \mathbf{r}_{12} v').$$

But, because the generators are tangents to the edge of regression, $\mathbf{R}'$ is parallel to $\mathbf{r}_1$ and therefore perpendicular to $\mathbf{r}_2$. Forming the scalar product with $\mathbf{r}_2$ we have

$$0 = Gv' + r\mathbf{r}_2 \bullet \mathbf{r}_{12} v' = v'\left(G + \tfrac{1}{2} r \frac{\partial G}{\partial u}\right),$$

the other terms vanishing in virtue of the relations $F=0$ and $E=1$. Hence, since v' is not zero,

$$\frac{1}{r} = -\frac{1}{2G}\frac{\partial G}{\partial u},$$

showing that $r = \rho$. Therefore the point Q on the edge of regression is the centre of geodesic curvature of the orthogonal trajectory of the geodesics.

57. Geodesic polar coordinates. An important particular case of the preceding is that in which the geodesics $v = \text{const.}$ are the singly infinite family of geodesics through a fixed point O, called the *pole*. Their orthogonal trajectories are the geodesic parallels $u = \text{const.}$, and we suppose u chosen so that $E = 1$. If we take the infinitesimal trajectory at the pole as the curve $u = 0$, u is the geodesic distance of a point from the pole. Hence the name *geodesic circles* given to the parallels $u = \text{const.}$ when the geodesics are concurrent. We may take v as the inclination of the geodesic at O to a fixed geodesic of reference OA. Then the position of any point P on the surface is determined by the geodesic through O on which it lies, and its distance u from O along that geodesic. These parameters u, v are called the *geodesic polar coordinates* of P. They are analogous to plane polar coordinates.

On a curve C drawn on the surface let P and Q be the consecutive points (u, v) and $(u + du, v + dv)$. Then dv is the angle at O

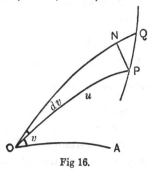

Fig 16.

between the geodesics OP and OQ. Let PN be an element of the geodesic circle through P, cutting OQ at N. Then $ON = OP$ and therefore $NQ = du$. And since the angle at N is a right angle,

$$NP^2 + du^2 = PQ^2 = ds^2$$
$$= du^2 + D^2 dv^2,$$

showing that $\qquad PN = Ddv.$

Hence if ψ is the angle NQP, at which the geodesic cuts the curve C,

$$\sin \psi = D\frac{dv}{ds}, \quad \cos \psi = \frac{du}{ds}, \quad \tan \psi = D\frac{dv}{du}.$$

And we may also notice that the area of the element of the surface bounded by the geodesics v, $v + dv$ and the geodesic circles u, $u + du$ is

$$dS = D\,du\,dv.$$

If the curve C is itself a geodesic, we may write the first of equations (27) for geodesics in the form

$$\frac{d}{ds}(\cos \psi) - D_1 \sin \psi \frac{dv}{ds} = 0,$$

or $\qquad\qquad \sin \psi\, d\psi + D_1 \sin \psi\, dv = 0.$

Hence, *for a geodesic,* $\qquad d\psi = - D_1 dv$ ....(28).

It is also important to notice that *at the pole D_1 has the value unity.* To see this we consider a *small* geodesic circle distant u from the pole. The element of a geodesic from the pole to the circle is practically straight, and the element of the geodesic circle is therefore $u\,dv$ to the first order. Thus near the origin

$$D = u + \text{terms of higher order,}$$

and therefore, at the pole, $D_1 = 1$.

58. Geodesic triangle. If dS is the area of an element of the surface at a point where the specific curvature is K, we call $K\,dS$ the second curvature of the element, and $\iint K\,dS$ taken over any portion of the surface is the whole second curvature of that portion. We shall now prove a theorem, due to Gauss, on the whole second curvature of a curvilinear triangle ABC bounded by geodesics. Such a triangle is called a *geodesic* triangle, and Gauss's theorem may be stated: *The whole second curvature of a geodesic triangle is equal to the excess of the sum of the angles of the triangle over two right angles.*

Let us choose geodesic polar coordinates with the vertex A as pole. Then the specific curvature is

$$K = -\frac{1}{D}\frac{\partial^2 D}{\partial u^2},$$

and the area of an element of the surface is $D\,du\,dv$. Consequently the whole second curvature Ω of the geodesic triangle is

$$\Omega = \iint K\,dS = -\iint \frac{\partial^2 D}{\partial u^2}\,du\,dv.$$

Integrate first with respect to u, from the pole A to the side BC.

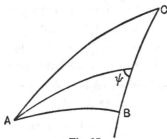

Fig. 17.

Then since at the pole D_1 is equal to unity, we find on integration

$$\Omega = \int (1 - D_1)\,dv,$$

where the integration with respect to v is along the side BC. But we have seen that, for a geodesic

$$-D_1\,dv = d\psi.$$

Hence our formula may be written

$$\Omega = \int dv + \int d\psi.$$

Now the first integral, taken from B to C, is equal to the angle A of the triangle. Also

$$\int d\psi = C - (\pi - B).$$

Hence the whole second curvature of the triangle is given by

$$\Omega = A + B + C - \pi \dots\dots\dots\dots\dots(29),$$

as required.

The specific curvature is positive, zero or negative according as
the surface is synclastic, developable or anticlastic. Consequently
$A + B + C$ is greater than π for a synclastic surface, equal to π for
a developable, and less than π for an anticlastic surface. When
the surface is a sphere Gauss's theorem is identical with Girard's
theorem on the area of a spherical triangle.

59. Theorem on parallels. An arbitrarily chosen family of
curves, $\phi(u, v) = \text{const.}$, does not in general constitute a system of
geodesic parallels. In order that they may do so, the function
$\phi(u, v)$ must satisfy a certain condition, which may be found as
follows. If the family of curves $\phi(u, v) = \text{const.}$ are geodesic
parallels to the family of geodesics $\psi(u, v) = \text{const.}$, the square of
the linear element can be expressed in the geodesic form

$$ds^2 = ed\phi^2 + D^2 d\psi^2,$$

where e is a function of ϕ only, and D a function of ϕ and ψ.
Equating two expressions for ds^2 we have the identity

$$E\,du^2 + 2F\,du\,dv + G\,dv^2 = e\,(\phi_1 du + \phi_2 dv)^2 + D^2\,(\psi_1 du + \psi_2 dv)^2,$$

and therefore
$$E = e\phi_1^2 + D^2\psi_1^2,$$
$$F = e\phi_1\phi_2 + D^2\psi_1\psi_2,$$
$$G = e\phi_2^2 + D^2\psi_2^2.$$

Consequently, eliminating ψ_1 and ψ_2, we must have

$$(E - e\phi_1^2)(G - e\phi_2^2) - (F - e\phi_1\phi_2)^2 = 0 \quad \ldots\ldots\ldots(a),$$

which is equivalent to

$$\frac{1}{H^2}(G\phi_1^2 - 2F\phi_1\phi_2 + E\phi_2^2) = \frac{1}{e} \quad \ldots\ldots\ldots(30).$$

Thus *in order that the family of curves $\phi(u, v) = \text{const.}$ may be a
family of geodesic parallels,*

$$(G\phi_1^2 - 2F\phi_1\phi_2 + E\phi_2^2)/H^2$$

must be a function of ϕ only, or a constant.
The condition is also *sufficient.* For

$$ds^2 - ed\phi^2 = (E - e\phi_1^2)\,du^2 + 2\,(F - e\phi_1\phi_2)\,du\,dv + (G - e\phi_2^2)\,dv^2$$

and this, regarded as a function of du and dv, is a perfect square,
in virtue of (a) being satisfied. We can therefore write it as $D^2 d\psi^2$,
so that

$$ds^2 = ed\phi^2 + D^2 d\psi^2,$$

proving the sufficiency of the condition. In order that ϕ may be the length of the geodesics measured from $\phi = 0$, it is necessary and sufficient that $e = 1$, that is

$$G\phi_1{}^2 - 2F\phi_1\phi_2 + E\phi_2{}^2 = H^2 \quad\ldots\ldots\ldots\ldots(30').$$

60. Geodesic ellipses and hyperbolas. Let two independent systems of geodesic parallels be taken as parametric curves, and let the parametric variables be chosen so that u and v are the actual geodesic distances of the point (u, v) from the particular

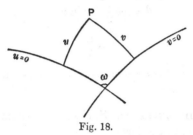

Fig. 18.

curves $u = 0$ and $v = 0$ (or from the poles in case the parallels are geodesic circles). Then by Art. 59, since the curves $u = $ const. and $v = $ const. are geodesic parallels for which $e = 1$, we have

$$E = G = H^2.$$

Hence, if ω is the angle between the parametric curves, it follows that

$$E = G = \frac{1}{\sin^2\omega}, \quad F = \frac{\cos\omega}{\sin^2\omega},$$

so that the square of the linear element is

$$ds^2 = \frac{du^2 + 2\cos\omega\,du\,dv + dv^2}{\sin^2\omega}\ldots\ldots\ldots\ldots(31).$$

And, conversely, when the linear element is of this form, the parametric curves are systems of geodesic parallels.

With this choice of parameters the locus of a point for which $u + v = $ const. is called a *geodesic ellipse*. Similarly the locus of a point for which $u - v = $ const. is a *geodesic hyperbola*. If we put

$$\bar{u} = \tfrac{1}{2}(u + v), \quad \bar{v} = \tfrac{1}{2}(u - v) \quad\ldots\ldots\ldots\ldots(32),$$

the above expression for ds^2 becomes

$$ds^2 = \frac{d\bar{u}^2}{\sin^2\dfrac{\omega}{2}} + \frac{d\bar{v}^2}{\cos^2\dfrac{\omega}{2}} \quad\ldots\ldots\ldots\ldots(33),$$

showing that the curves $\bar{u}=$ const. and $\bar{v}=$ const. are orthogonal. But these are geodesic ellipses and hyperbolas. Hence *a system of geodesic ellipses and the corresponding system of geodesic hyperbolas are orthogonal.* Conversely, whenever ds^2 is of the form (33), the substitution (32) reduces it to the form (31), showing that the parametric curves in (33) are geodesic ellipses and hyperbolas.

Further, if θ is the inclination of the curve $\bar{v}=$ const. to the curve $v=$ const., it follows from Art. 24 that

$$\cos\theta = \cos\frac{\omega}{2}, \quad \sin\theta = \sin\frac{\omega}{2},$$

and therefore

$$\theta = \frac{\omega}{2}.$$

Thus *the geodesic ellipses and hyperbolas bisect the angles between the corresponding systems of geodesic parallels.*

61. Liouville surfaces. Surfaces for which the linear element is reducible to the form

$$ds^2 = (U + V)(P\,du^2 + Q\,dv^2)\dots\dots\dots\dots(34),$$

in which U, P are functions of u alone, and V, Q are functions of v alone, were first studied by Liouville, and are called after him. The parametric curves clearly constitute an *isometric* system (Art. 39). It is also easy to show that they are a system of geodesic ellipses and hyperbolas. For if we change the parametric variables by the substitution

$$\sqrt{P}\,du = \frac{d\bar{u}}{\sqrt{U}}, \quad \sqrt{Q}\,dv = \frac{d\bar{v}}{\sqrt{V}},$$

the parametric curves are unaltered, and the linear element takes the form

$$ds^2 = (U + V)\left(\frac{d\bar{u}^2}{U} + \frac{d\bar{v}^2}{V}\right).$$

But this is of the form (33), where

$$\sin^2\frac{\omega}{2} = \frac{U}{U+V}, \quad \cos^2\frac{\omega}{2} = \frac{V}{U+V}.$$

Hence *the parametric curves are geodesic ellipses and hyperbolas.*

Liouville also showed that, when ds^2 has the form (34), a first integral of the differential equation of geodesics is given by

$$U\sin^2\theta - V\cos^2\theta = \text{const.} \dots\dots\dots\dots(35),$$

where θ is the inclination of the geodesic to the parametric curve $v = \text{const.}$ To prove this we observe that $F = 0$, while

$$E = (U + V) P, \qquad G = (U + V) Q,$$

so that $\quad E_1 = U_1 P + (U + V) P_1, \quad G_1 = U_1 Q,$

$$E_2 = V_2 P, \qquad G_2 = V_2 Q + (U + V) Q_2.$$

Taking the general equations (4) of geodesics, multiplying the first by $-2u' V$, the second by $2v' U$ and adding, we may arrange the result in the form

$$\frac{d}{ds}(UGv'^2 - VEu'^2) = u'^2 v' \{(U + V) E_2 - V_2 E\}$$

$$- u'v'^2 \{(U + V) G_1 - U_1 G\}.$$

Now the second member vanishes identically in virtue of the preceding relations. Hence

$$UGv'^2 - VEu'^2 = \text{const.,}$$

which, by Art. 24, is equivalent to

$$U \sin^2 \theta - V \cos^2 \theta = \text{const.}$$

as required.

EXAMPLES VIII

1. From formula (21) deduce the geodesic curvature of the curves $v = \text{const.}$ and $u = \text{const.}$

2. When the curves of an orthogonal system have constant geodesic curvature, the system is isometric.

3. If the curves of one family of an isometric system have constant geodesic curvature, so also have the curves of the other family.

4. Straight lines on a surface are the only asymptotic lines which are geodesics.

5. Find the geodesics of an ellipsoid of revolution.

6. If two families of geodesics cut at a constant angle, the surface is developable.

7. A curve is drawn on a cone, semi-vertical angle α, so as to cut the generators at a constant angle β. Prove that the torsion of its geodesic tangent is $\sin\beta \cos\beta/(R \tan a)$, where R is the distance from the vertex.

8. Prove that any curve is a geodesic on the surface generated by its binormals, and an asymptotic line on the surface generated by its principal normals.

9. Find the geodesics on the catenoid of revolution

$$u = c \cosh\frac{z}{c}.$$

10. If a geodesic on a surface of revolution cuts the meridians at a constant angle, the surface is a right cylinder.

11. If the principal normals of a curve intersect a fixed line, the curve is a geodesic on a surface of revolution, and the fixed line is the axis of the surface.

12. A curve for which κ/τ is constant is a geodesic on a cylinder: and a curve for which $\frac{d}{ds}(\tau/\kappa)$ is constant is a geodesic on a cone.

13. Show that the family of curves given by the differential equation $P\,du + Q\,dv = 0$ will constitute a system of geodesic parallels provided

$$\frac{\partial}{\partial u}\left(\frac{HQ}{\sqrt{EQ^2 - 2FPQ + GP^2}}\right) = \frac{\partial}{\partial v}\left(\frac{HP}{\sqrt{EQ^2 - 2FPQ + GP^2}}\right).$$

14. *If, on the geodesics through a point O, points be taken at equal geodesic distances from O, the locus of the points is an orthogonal trajectory of the geodesics.*

Let the geodesics through the pole O be taken as the curves $v = $ const., and let u denote the geodesic distance measured from the pole. We have to show that the parametric curves are orthogonal. Since the element of arc of a geodesic is du, it follows that $E = 1$. Also since the curves $v = $ const. are geodesics, $\lambda = 0$. Hence $F_1 = 0$, so that F is a function of v alone. Now, at the pole, $\mathbf{r}_2$ is zero, and therefore $\mathbf{r}_1 \cdot \mathbf{r}_2 \equiv F$ vanishes at the pole. But F is independent of u; and therefore it vanishes along any geodesic. Thus F vanishes identically, and the parametric curves are orthogonal.

15. *If, on the geodesics which cut a given curve C orthogonally, points be taken at equal geodesic distances from C, the locus of the points is an orthogonal trajectory of the geodesics.*

16. Necessary and sufficient conditions that a system of geodesic coordinates be polar are that $\sqrt{G}$ vanishes with u, and $\partial\sqrt{G}/\partial u = 1$ when $u = 0$.

17. Two points A, B on the surface are joined by a fixed curve C_0 and a variable curve C, enclosing between them a portion of the surface of constant area. Prove that the length of C is least when its geodesic curvature is constant.

18. If in the previous example the length of C is constant, prove that the area enclosed is greatest when the geodesic curvature of C is constant.

19. If the tangent to a geodesic is inclined at a constant angle to a fixed direction, the normal to the surface along the geodesic is everywhere perpendicular to the fixed direction.

20. Two surfaces touch each other along a curve. If the curve is a geodesic on one surface, it is a geodesic on the other also.

21. The ratio of the curvature to the torsion of a geodesic on a developable surface is equal to the tangent of the inclination of the curve to the corresponding generating line.

22. If a geodesic on a developable surface is a plane curve, it is one of the generators, or else the surface is a cylinder.

23. If a geodesic on a surface lie on a sphere, the radius of curvature of the geodesic is equal to the perpendicular from the centre of the sphere on the tangent plane to the surface.

24. The locus of the centre of geodesic curvature of a line of curvature is an evolute of the latter.

25. The orthogonal trajectories of the helices on a helicoid are geodesics.

26. The meridians of a ruled helicoid are geodesics.

27. If the curve

$$x = f(u) \cos u, \quad y = f(u) \sin u, \quad z = -\frac{1}{c} \int f^2(u) \, du$$

is given a helicoidal motion of pitch $2\pi c$ about the z-axis, the various positions of the curve are orthogonal trajectories of the helices, and also geodesics on the surface.

CHAPTER VII

QUADRIC SURFACES. RULED SURFACES

QUADRIC SURFACES

62. Central quadrics. The equation of a central quadric surface, referred to its principal axes, is of the form

$$\frac{x^2}{a} + \frac{y^2}{b} + \frac{z^2}{c} = 1 \quad \dots\dots\dots\dots\dots(1),$$

in which we may assume $a > b > c$. The quadrics confocal with this are given by

$$\frac{x^2}{a+\lambda} + \frac{y^2}{b+\lambda} + \frac{z^2}{c+\lambda} = 1 \quad \dots\dots\dots\dots(2),$$

for different values of λ. At points common to the two surfaces (1) and (2) we have

$$\phi(\lambda) \equiv (a+\lambda)(b+\lambda)(c+\lambda) - \Sigma x^2 (b+\lambda)(c+\lambda) = 0.$$

We may regard this as an equation for determining the values of λ corresponding to the confocals which pass through a given point (x, y, z) on the surface (1). It is a cubic equation, one root of which is obviously zero. Let the other two roots be denoted by u, v. Then, because the coefficient of λ^3 is unity, $\phi(\lambda)$ is identically equal to the product $\lambda(\lambda - u)(\lambda - v)$; that is

$$\lambda(\lambda - u)(\lambda - v) \equiv (a+\lambda)(b+\lambda)(c+\lambda) - \Sigma x^2 (b+\lambda)(c+\lambda).$$

If in this identity we give λ the values $-a, -b, -c$ in succession, we find

$$\left. \begin{aligned} x^2 &= \frac{a(a+u)(a+v)}{(a-b)(a-c)} \\[2mm] y^2 &= \frac{b(b+u)(b+v)}{(b-a)(b-c)} \\[2mm] z^2 &= \frac{c(c+u)(c+v)}{(c-a)(c-b)} \end{aligned} \right\} \dots\dots\dots\dots\dots(3).$$

Thus the coordinates of a point on the quadric (1) are expressible in terms of the parameters u, v of the two confocals passing through that point. We take these for parametric variables on the surface. It follows from (3) that, for given values of u and v, there are eight points on the surface, one in each octant, symmetrically situated with respect to the coordinate planes.

In the case of an *ellipsoid*, a, b, c are all positive. Hence $\phi(-c)$ is negative, $\phi(-b)$ positive, and $\phi(-a)$ negative. Therefore, if u is greater than v, we have

$$-c > u > -b, \quad -b > v > -a.$$

The values of u and v are thus negative, and are separated by $-b$.

For an *hyperboloid of one sheet* c is negative, so that $\phi(\infty)$ is positive, $\phi(-c)$ negative, $\phi(-b)$ positive and $\phi(-a)$ negative. Therefore

$$u > -c, \quad -b > v > -a.$$

Consequently u is positive and v negative, the root between $-c$ and $-b$ being the zero root.

For an *hyperboloid of two sheets* both b and c are negative. Hence $\phi(\infty)$ is positive, $\phi(-c)$ negative and $\phi(-b)$ positive, so that the non-zero roots are both positive and such that

$$u > -c, \quad -c > v > -b.$$

Thus both parameters are positive, and the values of u and v are separated by $-c$. In all cases one of the three surfaces through (x, y, z) is an ellipsoid, one an hyperboloid of one sheet, and one an hyperboloid of two sheets.

Any parametric curve $v = $ const. on the quadric (1) is the curve of intersection of the surface with the confocal of parameter equal to this constant v. Similarly any curve $u = $ const. is the line of intersection of the surface with the confocal of parameter equal to this constant u.

63. Fundamental magnitudes. If r is the distance of the point (x, y, z) from the centre of the quadric, and p the length of the central perpendicular on the tangent plane at (x, y, z), we have

$$r^2 = x^2 + y^2 + z^2 = (a+b+c) + (u+v)$$

and

$$\frac{1}{p^2} = \frac{x^2}{a^2} + \frac{y^2}{b^2} + \frac{z^2}{c^2} = \frac{uv}{abc} \qquad \Bigg\} \quad \ldots\ldots(4).$$

Also on calculating the partial derivatives x_1, x_2, etc., we find

$$E = x_1^2 + y_1^2 + z_1^2 = \frac{u(u-v)}{4(a+u)(b+u)(c+u)}$$

$$F = x_1 x_2 + y_1 y_2 + z_1 z_2 = 0 \qquad \Bigg\} \quad \ldots\ldots(5).$$

$$G = x_2^2 + y_2^2 + z_2^2 = \frac{v(v-u)}{4(a+v)(b+v)(c+v)}$$

The normal has the direction of the vector $\left(\dfrac{x}{a}, \dfrac{y}{b}, \dfrac{z}{c}\right)$; and since the square of this vector is equal to $1/p^2$, the unit normal is

$$\mathbf{n} = \left(\frac{px}{a}, \frac{py}{b}, \frac{pz}{c}\right)$$

$$= \left(\sqrt{\frac{bc}{uv}\frac{(a+u)(a+v)}{(a-b)(a-c)}}, \quad \sqrt{\frac{ca}{uv}\frac{(b+u)(b+v)}{(b-c)(b-a)}}, \right.$$

$$\left. \sqrt{\frac{ab}{uv}\frac{(c+u)(c+v)}{(c-a)(c-b)}}\right).$$

The second order magnitudes are therefore

$$\left. \begin{aligned}
L &= \mathbf{n} \cdot \mathbf{r}_{11} = \frac{1}{4}\sqrt{\frac{abc}{uv}}\frac{(u-v)}{(a+u)(b+u)(c+u)} \\
M &= \mathbf{n} \cdot \mathbf{r}_{12} = 0 \\
N &= \mathbf{n} \cdot \mathbf{r}_{22} = \frac{1}{4}\sqrt{\frac{abc}{uv}}\frac{(v-u)}{(a+v)(b+v)(c+v)}
\end{aligned} \right\} \quad \dots \dots (6).$$

Since then $F = 0$ and $M = 0$ *the parametric curves are lines of curvature.* That is to say, the lines of curvature on a central quadric are the curves in which it is cut by the confocals of different species. The principal curvatures are then given by

$$\left. \begin{aligned}
\kappa_a &= \frac{L}{E} = \frac{1}{u}\sqrt{\frac{abc}{uv}} \\
\kappa_b &= \frac{N}{G} = \frac{1}{v}\sqrt{\frac{abc}{uv}}
\end{aligned} \right\} \quad \dots \dots \dots \dots (7).$$

Thus, *along a line of curvature, the principal curvature varies as the cube of the other principal curvature.* The first curvature is

$$J = \kappa_a + \kappa_b = (u+v)\sqrt{\frac{abc}{(uv)^3}},$$

and the specific curvature

$$K = \kappa_a \kappa_b = \frac{abc}{u^2 v^2} \quad \dots \dots \dots \dots (8).$$

Therefore on the ellipsoid or the hyperboloid of two sheets the specific curvature is positive at all points; but on the hyperboloid of one sheet it is negative everywhere. Moreover

$$p^4 = abc\,K \quad \dots \dots \dots \dots (9).$$

Hence *at all points of a curve, at which the specific curvature is constant, the tangent plane is at a constant distance from the centre.*

At an *umbilic* κ_a is equal to κ_b, and therefore $u = v$. If the surface is an ellipsoid the values of u and v are separated by $-b$. Hence at an umbilic they must have the common value $-b$. The umbilici are therefore

$$x = \pm \sqrt{\frac{a(a-b)}{(a-c)}}, \quad y = 0, \quad z = \pm \sqrt{\frac{c(b-c)}{(a-c)}}.$$

The four umbilici thus lie on the coordinate plane containing the greatest and least axes, and are symmetrically situated with respect to those axes.

On the hyperboloid of two sheets the values of the parameters are separated by $-c$. Hence at an umbilic $u = v = -c$, and the umbilici are

$$x = \pm \sqrt{\frac{a(a-c)}{a-b}}, \quad y = \pm \sqrt{\frac{b(b-c)}{(b-a)}}, \quad z = 0.$$

On the hyperboloid of one sheet the umbilici are imaginary, for u and v have no common value.

The differential equation of the *asymptotic lines* on a surface is

$$L du^2 + 2M du dv + N dv^2 = 0.$$

Hence on the quadric (1) they are given by

$$\frac{du}{\sqrt{(a+u)(b+u)(c+u)}} = \pm \frac{dv}{\sqrt{(a+v)(b+v)(c+v)}}.$$

64. Geodesics. On using the values of E, F, G given in (5) we see that the square of the linear element takes the form

$$ds^2 = (u - v)(U du^2 - V dv^2),$$

where U is a function of u alone, and V a function of v alone. Central quadrics thus belong to the class of surfaces called Liouville surfaces (Art. 61). Consequently *the lines of curvature, being parametric curves, are isometric and constitute a system of geodesic ellipses and hyperbolas.* Moreover a first integral of the differential equation of geodesics on the quadric is given by

$$u \sin^2 \theta + v \cos^2 \theta = k \quad \ldots\ldots\ldots\ldots(10),$$

where k is constant, and θ the angle at which the geodesic cuts the curve $v = $ const. The value of k is constant on any one geodesic, but changes from one geodesic to another. If the geodesic touches the parametric curve $v = h$, then $\cos \theta = 1$ at the point of contact, and therefore $k = h$. Similarly if it touches the curve $u = h$, $\sin \theta = 1$ at the point of contact, and again $k = h$. Thus *k has the same value for all those geodesics which touch the same line of curvature.* On the ellipsoid k is negative for all geodesics because both parameters are negative. On the hyperboloid of two sheets k is positive because u and v are both positive.

Further, on writing (10) in the form

$$(u - k) \sin^2 \theta + (v - k) \cos^2 \theta = 0,$$

we see that, for all geodesics through a given point (u, v), the constant k is intermediate in value between u and v; and, for a given value of k within this interval, there are two geodesics, and these are equally inclined to the lines of curvature.

At an umbilic the parametric values u, v are equal; and therefore, for all geodesics through an umbilic, k has the same value, k_0 say, which is $-b$ for an ellipsoid and $-c$ for an hyperboloid of two sheets. The equation for the *umbilical geodesics* is then

$$(u - k_0) \sin^2 \theta + (v - k_0) \cos^2 \theta = 0.$$

Thus, *through each point P on a central quadric with real umbilics there pass two umbilical geodesics, and these are equally inclined to the lines of curvature through the point.* If then the point P is

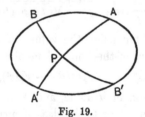

Fig. 19.

joined by geodesics to the four umbilics, those drawn to opposite umbilics A, A' or B, B' must be continuations of each other. Thus two opposite umbilics are joined by an infinite number of geodesics, no two of which intersect again.

Moreover, since the geodesics joining P to two consecutive umbilics A, B are equally inclined to the lines of curvature at P, it follows that A, B are foci of the geodesic ellipses and hyperbolas formed by the lines of curvature. Of the two lines of curvature through P, that one is a geodesic ellipse with respect to A and B

which bisects externally the angle APB, while the other one, which bisects the angle internally, is a geodesic hyperbola. If, however, A' and B are taken as foci, the former is a geodesic hyperbola and the latter a geodesic ellipse. Thus, in the case of the former,

$$PA + PB = \text{const.},$$
$$PA' - PB = \text{const.},$$

so that $\qquad PA + PA' = \text{const.}$

But P is any point on the surface. Hence *all the geodesics joining two opposite umbilics are of equal length.*

65. Other properties. On using the values of the principal curvatures given in (7) we deduce from Euler's theorem that the curvature of a geodesic, being the normal curvature of the surface in that direction, is given by

$$\kappa_n = \frac{1}{u} \sqrt{\frac{\overline{abc}}{uv}} \cos^2 \theta + \frac{1}{v} \sqrt{\frac{\overline{abc}}{uv}} \sin^2 \theta,$$

so that $\qquad \kappa_n = \dfrac{kp^3}{abc}$ ....(11).

Hence *along any one geodesic the normal curvature varies as the cube of the central perpendicular on the tangent plane.* The same is also true of a line of curvature. For, at any point, κ_n and p have the same values for this curve as for the geodesic tangent, and all geodesic tangents to a line of curvature have the same k.

Again, consider the semi-diameter D of the quadric parallel to the tangent to the geodesic at the point (x, y, z). The *unit* tangent $\mathbf{r}'$ to the geodesic is (x', y', z') and therefore

$$\frac{1}{D^2} = \frac{x'^2}{a} + \frac{y'^2}{b} + \frac{z'^2}{c} \quad \dots\dots\dots\dots\dots\dots\dots(\alpha),$$

while, for any direction on the surface,

$$\frac{xx'}{a} + \frac{yy'}{b} + \frac{zz'}{c} = 0 \quad \dots\dots\dots\dots\dots\dots(\beta).$$

Along a geodesic $\mathbf{r}'' = \kappa_n \mathbf{n}$; and therefore, by differentiating the identity $\mathbf{r}' \cdot \mathbf{n} = 0$, we have

$$\kappa_n = \mathbf{r}'' \cdot \mathbf{n} = - \mathbf{r}' \cdot \mathbf{n}'.$$

Now
$$\mathbf{n} = p\left(\frac{x}{a}, \frac{y}{b}, \frac{z}{c}\right),$$

and therefore
$$\mathbf{n}' = p\left(\frac{x'}{a}, \frac{y'}{b}, \frac{z'}{c}\right) + p'\left(\frac{x}{a}, \frac{y}{b}, \frac{z}{c}\right).$$

Thus on forming the scalar product $\mathbf{r}' \cdot \mathbf{n}'$ we find, in virtue of (α) and (β),

$$\kappa_n = -p\left(\frac{x'^2}{a} + \frac{y'^2}{b} + \frac{z'^2}{c}\right) = -\frac{p}{D^2} \quad \ldots\ldots\ldots\ldots(12).$$

On substituting the value of κ_n given in (11) we have

$$kp^2 D^2 = -abc \ldots\ldots\ldots\ldots\ldots\ldots(13),$$

or
$$kD^2 = -uv.$$

From (13) it follows that pD is constant along a geodesic. The same is also true of a line of curvature. For p and D are the same for the line of curvature as for its geodesic tangent; while k has the same value for all geodesic tangents to a line of curvature. Thus we have Joachimsthal's theorem : *Along a geodesic or a line of curvature on a central quadric the product of the semi-diameter of the quadric parallel to the tangent to the curve and the central perpendicular to the tangent plane is constant.*

Formula (12) shows that $\kappa_n D^2 = -p$. Now p is the same for all directions at a point; and therefore, if ρ is the reciprocal of κ_n, D varies as $\sqrt{\rho}$. Hence, by Art. 32, the indicatrix at any point of a central quadric is similar and similarly situated to the parallel central section.

Ex. 1. Show that, along a geodesic or a line of curvature, κ_n varies inversely as D^3.

Ex. 2. For all umbilical geodesics on the quadric (1) $p^2 D^2 = ac$.

Ex. 3. The constant pD has the same value for all geodesics that touch the same line of curvature.

Ex. 4. Two geodesic tangents to a line of curvature are equally inclined to the lines of curvature through their point of intersection.

Ex. 5. The geodesic distance between two opposite umbilics on an ellipsoid is one half the circumference of the principal section through the umbilics.

Ex. 6. All geodesics through an umbilic on an ellipsoid pass through the opposite umbilic.

*66. Paraboloids. The equation of a paraboloid may be expressed in the form

$$\frac{x^2}{a} + \frac{y^2}{b} = 4z \quad \ldots\ldots\ldots\ldots(14),$$

in which we may assume that b is positive and greater than a. The paraboloids confocal with this are given by

$$\frac{x^2}{a-\lambda} + \frac{y^2}{b-\lambda} = 4(z-\lambda),$$

for different values of λ. The values of λ for the confocals through a given point (x, y, z) on the original surface are given by

$$\phi(\lambda) \equiv x^2(b-\lambda) + y^2(a-\lambda) - 4(z-\lambda)(a-\lambda)(b-\lambda) = 0.$$

One root of this cubic is zero : let the other roots be denoted by u, v of which u is the greater. Then, because the coefficient of λ^3 in the last equation is 4, we have the identity

$$4\lambda(\lambda - u)(\lambda - v) \equiv x^2(b-\lambda) + y^2(a-\lambda) - 4(z-\lambda)(a-\lambda)(b-\lambda).$$

If in this we give λ the values a and b successively, we find

$$\left. \begin{array}{l} x^2 = \dfrac{4a(a-u)(a-v)}{b-a} \\[2mm] y^2 = \dfrac{4b(b-u)(b-v)}{a-b} \end{array} \right\} \quad \ldots\ldots\ldots\ldots(15).$$

and therefore by (14) $z = u + v - a - b$

We may take u, v for parameters on the paraboloid; and for given values of the parameters there are four points on the surface, symmetrically situated with respect to the coordinate planes $x = 0$ and $y = 0$.

For the *elliptic paraboloid* a is positive as well as b. Hence since $\phi(\infty)$ is positive, $\phi(b)$ negative and $\phi(a)$ positive, it follows that u and v are both positive, and are separated by the value b. For the *hyperbolic paraboloid* a is negative. The zero root of $\phi(\lambda)$ lies between a and b, so that $u > b$ and $v < a$.

The derivatives of x, y, z are easily calculated from (15), and the first order magnitudes found to be

$$\left. \begin{array}{l} E = \Sigma x_1{}^2 = \dfrac{u(u-v)}{(a-u)(b-u)} \\[2mm] F = \Sigma x_1 x_2 = 0 \\[2mm] G = \Sigma x_2{}^2 = \dfrac{v(v-u)}{(a-v)(b-v)} \end{array} \right\} \quad \ldots\ldots\ldots\ldots(16).$$

The normal to the paraboloid has the direction of the vector $\left(-\dfrac{x}{a}, -\dfrac{y}{b}, 2\right)$; and the *unit* vector in this direction, expressed in terms of the parameters, is

$$\mathbf{n} = \left(-\sqrt{\frac{b\,(a-u)\,(a-v)}{(b-a)\,uv}}, \; -\sqrt{\frac{a\,(b-u)\,(b-v)}{(a-b)\,uv}}, \; \sqrt{\frac{ab}{uv}}\right).$$

The second order magnitudes are therefore

$$L = \mathbf{n} \cdot \mathbf{r}_{11} = \frac{u-v}{2\,(a-u)\,(b-u)}\sqrt{\frac{ab}{uv}},$$

$$M = \mathbf{n} \cdot \mathbf{r}_{12} = 0,$$

$$N = \mathbf{n} \cdot \mathbf{r}_{22} = \frac{v-u}{2\,(a-v)\,(b-v)}\sqrt{\frac{ab}{uv}}.$$

Since then F and M both vanish identically, the parametric curves are lines of curvature. That is to say, *the lines of curvature on a paraboloid are the curves in which it is cut by the confocals.* The principal curvatures are then given by

$$\left.\begin{aligned} \kappa_a &= \frac{L}{E} = \frac{1}{2u}\sqrt{\frac{ab}{uv}} \\[2mm] \kappa_b &= \frac{N}{G} = \frac{1}{2v}\sqrt{\frac{ab}{uv}} \end{aligned}\right\} \quad \ldots\ldots\ldots\ldots(17),$$

and the specific curvature is

$$K = \kappa_a \kappa_b = \frac{ab}{4u^2 v^2}.$$

The length p of the perpendicular from the vertex to the tangent plane at the point (x, y, z) is easily found to be

$$p = z\sqrt{\frac{ab}{uv}}.$$

Hence *the quotient p/z is constant along a curve on which the specific curvature of the surface is constant.*

The *umbilici* are given by $\kappa_a = \kappa_b$, which requires $u = v$. This is possible only on an elliptic paraboloid; and the common parameter value is then equal to b. The umbilici on an elliptic paraboloid are therefore given by

$$x = \pm\, 2\sqrt{a\,(b-a)}, \quad y = 0, \quad z = b - a.$$

At these points the principal curvatures become equal to $\dfrac{1}{2}\sqrt{\dfrac{a}{b^3}}.$

In virtue of (16) the square of the linear element has Liouville' form

$$ds^2 = (u - v)(U du^2 - V dv^2),$$

so that *the lines of curvature are isometric and constitute a system of geodesic ellipses and hyperbolas.* A first integral of the differential equation of geodesics is given by

$$u \sin^2 \theta + v \cos^2 \theta = k,$$

as in the case of the central quadrics; and the direct consequences of this equation, which do not depend upon the existence of a centre, are true of the paraboloids also (Art. 64).

The curvature of a geodesic, being the normal curvature in its direction, is given by

$$\kappa_n = \kappa_a \cos^2 \theta + \kappa_b \sin^2 \theta$$

$$= \frac{k}{2} \sqrt{\frac{ab}{u^3 v^3}}.$$

Hence, *along a geodesic or a line of curvature on a paraboloid, the quotient* $\kappa_n z^3 / p^3$ *is constant.*

EXAMPLES IX

1. The points at which two geodesic tangents to a given line of curvature on a quadric cut orthogonally lie on the surface of a sphere.

Let k be the parametric constant for the line of curvature. Then for a geodesic tangent

$$u \sin^2 \theta + v \cos^2 \theta = k.$$

Where this intersects another geodesic tangent at right angles we have

$$u \sin^2 \left(\frac{\pi}{2} + \theta\right) + v \cos^2 \left(\frac{\pi}{2} + \theta\right) = k.$$

Hence, by addition, $u + v = 2k,$

and therefore by (4) $x^2 + y^2 + z^2 = a + b + c + 2k,$

which proves the theorem.

2. The points at which the geodesic tangents to two different lines of curvature cut orthogonally lie on a sphere.

3. If a geodesic is equally inclined to the lines of curvature at every point along it, prove that the sum of the principal curvatures varies as the cube of the central perpendicular to the tangent plane along the geodesic.

4. The intersection of a tangent to a given geodesic on a central conicoid with a tangent plane to which it is perpendicular lies on a sphere.

5. Prove that the differential equation of geodesics on an ellipsoid may be expressed

$$\frac{du}{dv} = \pm \sqrt{\frac{v(a+u)(b+u)(c+u)(u-k)}{u(a+v)(b+v)(c+v)(v-k)}}.$$

6. The length of an element of an umbilical geodesic on an ellipsoid is

$$\tfrac{1}{2}du \sqrt{\frac{u}{(a+u)(c+u)}} + \tfrac{1}{2}dv \sqrt{\frac{v}{(a+v)(c+v)}}.$$

7. For any curve on a quadric the product of the geodesic curvature and the torsion of the geodesic tangent is equal to

$$\tfrac{1}{2}p^3 \frac{d}{ds}\left(\frac{1}{pD}\right)^2,$$

with the notation of Art. 65.

8. The asymptotic lines on any quadric are straight lines.

9. Find the area of an element of a quadric bounded by four lines of curvature.

10. A geodesic is drawn from an umbilic on an ellipsoid to the extremity of the mean axis. Show that its torsion at the latter point is

$$\frac{1}{ac} \sqrt{b(a-b)(b-c)}.$$

11. Find the geodesic curvature of the lines of curvature on a central quadric.

12. Find the tangent of the angle between the umbilical geodesics through the point (x, y, z) on the ellipsoid (1).

13. The specific curvature at every point of the elliptic paraboloid $ax^2 + by^2 = 2z$ where it is cut by the cylinder $a^2x^2 + b^2y^2 = 1$ is $\tfrac{1}{4}ab$.

14. The specific curvature at any point of a paraboloid varies as p^4/z^4, with the notation of Art. 66.

15. Writing ds^2 for a quadric in the form

$$ds^2 = (u-v)(U du^2 - V dv^2),$$

prove that the quantities l, λ, etc. are given by

$$l = \frac{1}{2}\left(\frac{1}{u-v} + \frac{U'}{U}\right), \quad m = -\frac{1}{2(u-v)}, \quad n = \frac{V}{2(u-v)U},$$

$$\lambda = \frac{U}{2(v-u)V}, \quad \mu = -\frac{1}{2(v-u)}, \quad \nu = \frac{1}{2}\left(\frac{1}{v-u} + \frac{V'}{V}\right).$$

Hence write down the (single) differential equation of geodesics.

16. Show that the coordinates of the centres of curvature for a central quadric are, for one principal direction,

$$\sqrt{\frac{(a+u)^3(a+v)}{a(a-b)(a-c)}}, \quad \sqrt{\frac{(b+u)^3(b+v)}{b(b-c)(b-a)}}, \quad \sqrt{\frac{(c+u)^3(c+v)}{c(c-a)(c-b)}},$$

and, for the other, similar expressions obtained by interchanging u and v. Hence show that the two sheets of the centro-surface are identical.

17. Show that the equation of a paraboloid, $ax^2 + by^2 = 2z$, is satisfied identically by the substitution

$$x^2 = \frac{a-b}{b} uv, \quad y^2 = \frac{b-a}{ab^2}(1+au)(1+av), \quad z = \frac{b-a}{2ab}(1+au+av).$$

Taking u, v as parameters for the surface, show that

$$E = \frac{(a-b)}{4b^2} \frac{(u-v) P^2}{u(1+au)},$$

$$F = 0,$$

$$G = \frac{(b-a)}{4b^2} \frac{(u-v) Q^2}{v(1+av)},$$

where $\quad P^2 = a(a-b)u - b, \quad Q^2 = a(a-b)v - b.$

Prove also that the unit normal to the surface is

$$\mathbf{n} = \frac{1}{PQ}(\sqrt{a^3(a-b)uv}, \quad \sqrt{b(b-a)(1+au)(1+av)}, \quad -\sqrt{ab}),$$

and the second order magnitudes

$$L = \frac{1}{4PQ}\sqrt{\frac{a^3}{b}} \frac{(a-b)(u-v)}{u(1+au)},$$

$$M = 0,$$

$$N = -\frac{1}{4PQ}\sqrt{\frac{a^3}{b}} \frac{(a-b)(u-v)}{v(1+av)}.$$

Hence deduce all the results of Art. 66.

Ruled Surfaces

67. Skew surface or scroll. A *ruled surface* is one that can be generated by the motion of a straight line. The infinitude of straight lines which thus lie on the surface are called its generators. We have already considered a particular class of ruled surfaces called developable surfaces or *torses*. These are characterised by the properties that consecutive generators intersect, that all the generators are tangents to a curve called the edge of regression, that the tangent plane to the surface is the same at all points of a given generator, and that the specific curvature of the surface is identically zero. Ruled surfaces in general, however, do not possess these properties. Those which are not developable are called *skew surfaces* or *scrolls*. It is skew surfaces particularly that we shall now consider.

On the given ruled surface let any curve be drawn cutting all the generators, and let it be taken as a curve of reference called the *directrix*. The position vector $\mathbf{r}_0$ of a current point P_0 on the directrix is a function of the arc-length s of this curve, measured from a fixed point on it. The position vector $\mathbf{r}$ of any point P on the surface is then given by

$$\mathbf{r} = \mathbf{r}_0 + u\mathbf{d} \qquad \dots \dots \dots \dots \dots (18),$$

where $\mathbf{d}$ is the *unit* vector parallel to the generator through P,

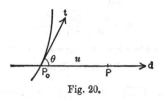

Fig. 20.

and u the distance of P from the directrix in the direction of $\mathbf{d}$. The quantities u, s will be taken as parameters for the surface. The parametric curves $s = $ const. are the generators. The unit tangent $\mathbf{t}$ to the directrix is equal to $\mathbf{r}_0'$; and the angle θ at which a generator cuts the directrix is given by

$$\cos \theta = \mathbf{d} \cdot \mathbf{t} \qquad \dots \dots \dots \dots \dots (19).$$

The square of the linear element of the surface follows from (18). For

$$d\mathbf{r} = \mathbf{d}\, du + (\mathbf{t} + u\mathbf{d}')\, ds$$

and therefore on squaring, and writing

$$a^2 = \mathbf{d}'^2, \quad b = \mathbf{t} \cdot \mathbf{d}' \qquad \dots \dots \dots \dots (20),$$

we have

$$E\, du^2 + 2F\, du\, ds + G\, ds^2 = d\mathbf{r}^2$$
$$= du^2 + 2 \cos \theta\, du\, ds + (a^2 u^2 + 2bu + 1)\, ds^2 \dots (21).$$

68. Consecutive generators. Consider consecutive generators through the points $\mathbf{r}_0$ and $\mathbf{r}_0 + \mathbf{t}\, ds$ on the directrix, and let their directions be those of the unit vectors $\mathbf{d}$ and $\mathbf{d} + \mathbf{d}'\, ds$. If their mutual moment is positive the ruled surface is said to be *right-handed*; if it is negative the surface is *left-handed*. This mutual moment is the scalar moment about either generator of a unit vector localised in the other. If then we take the unit vector $\mathbf{d} + \mathbf{d}'\, ds$ localised in the second generator, its vector moment about the point $\mathbf{r}_0$ is $(\mathbf{t}\,ds) \times (\mathbf{d} + \mathbf{d}'\,ds)$. Hence its scalar moment about the first generator is

$$(\mathbf{t}\,ds) \times (\mathbf{d} + \mathbf{d}'\,ds) \cdot \mathbf{d} = [\mathbf{t},\, \mathbf{d}',\, \mathbf{d}]\,ds^2.$$

The surface is therefore right-handed or left-handed according as the scalar triple product

$$D = [\mathbf{t},\, \mathbf{d}',\, \mathbf{d}] \quad \ldots\ldots\ldots\ldots\ldots(22)$$

is positive or negative.

The common perpendicular to the two generators is parallel to the vector $(\mathbf{d} + \mathbf{d}'\,ds) \times \mathbf{d}$, and therefore to the vector $\mathbf{d}' \times \mathbf{d}$. The unit vector in this direction is $\dfrac{1}{a}\,\mathbf{d}' \times \mathbf{d}$, because $\mathbf{d}'$ and $\mathbf{d}$ are at right angles, and their moduli are a and unity respectively. In the case of a right-handed surface this vector makes an acute angle with $\mathbf{t}$. The *shortest distance* between the consecutive generators is the projection of the arc-element $\mathbf{t}\,ds$ on the common perpendicular, and is therefore equal to

$$(\mathbf{t}\,ds) \cdot \left(\frac{1}{a}\,\mathbf{d}' \times \mathbf{d}\right) = \frac{ds}{a}\,[\mathbf{t},\, \mathbf{d}',\, \mathbf{d}] = \frac{D}{a}\,ds.$$

Hence *the necessary and sufficient condition that the surface be developable is* $[\mathbf{t},\, \mathbf{d}',\, \mathbf{d}] \equiv 0$.

This condition may be expressed differently. For

$$[\mathbf{t},\, \mathbf{d}',\, \mathbf{d}]^2 = \begin{vmatrix} \mathbf{t}^2 & \mathbf{t}\cdot\mathbf{d}' & \mathbf{t}\cdot\mathbf{d} \\ \mathbf{d}'\cdot\mathbf{t} & \mathbf{d}'^2 & \mathbf{d}'\cdot\mathbf{d} \\ \mathbf{d}\cdot\mathbf{t} & \mathbf{d}\cdot\mathbf{d}' & \mathbf{d}^2 \end{vmatrix} = \begin{vmatrix} 1 & b & \cos\theta \\ b & a^2 & 0 \\ \cos\theta & 0 & 1 \end{vmatrix}$$

$$= a^2 \sin^2\theta - b^2.$$

Hence $\qquad\qquad [\mathbf{t},\, \mathbf{d}',\, \mathbf{d}] = \pm \sqrt{a^2 \sin^2\theta - b^2} \quad \ldots\ldots\ldots\ldots(23),$

the positive or negative sign being taken according as the surface is right-handed or left-handed. Thus *the condition for a developable surface is*

$$b^2 = a^2 \sin^2\theta.$$

The mutual moment of two given generators and their shortest distance apart are clearly independent of the curve chosen as directrix. Hence, in the case of two consecutive generators, the quantities $D\,ds^2$ and $\dfrac{1}{a}\,D\,ds$ do not change with the directrix. The quotient of the square of the second by the first is then likewise invariant. But this quotient, being equal to D/a^2, is independent of ds, and therefore depends only on the particular generator $\mathbf{d}$ chosen. It is called the *parameter of distribution* for that generator, and has the same sign as D. Denoting it by β we have

$$\beta = \frac{D}{a^2} \quad\dots\dots\dots\dots\dots\dots(24).$$

69. Line of striction. The foot of the common perpendicular to a generator and the consecutive generator is called the *central point* of the generator; and the locus of the central points of all the generators is the *line of striction* of the surface. To find the distance $\bar{u}$ from the directrix to the central point of the generator $\mathbf{d}$, we first prove that the tangent to the line of striction is perpendicular to $\mathbf{d}'$. This may be done as follows. Consider three

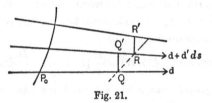

Fig. 21.

consecutive generators. Let QQ' be the element of the common perpendicular to the first and second intercepted between them, and RR' the intercept of the common perpendicular to the second and third. Then the vector QR is the sum of the vectors QQ' and $Q'R$. But QQ' is parallel to $\mathbf{d}' \times \mathbf{d}$ and is therefore perpendicular to $\mathbf{d}'$. Further

$$Q'R = \eta\,(\mathbf{d} + \mathbf{d}'\,ds),$$

where η is a small quantity of the first order. Forming the scalar product of this vector with $\mathbf{d}'$ we have for its value

$$\eta\,(\mathbf{d} + \mathbf{d}'\,ds)\cdot\mathbf{d}' = \eta a^2 ds,$$

which is of the second order. Hence in the limit, as the three generators tend to coincidence, QR is perpendicular to $\mathbf{d}'$. But the limiting direction of QR is that of the tangent to the line of striction. Hence this tangent is perpendicular to $\mathbf{d}'$.

Now if $\bar{u}$ is the distance of the central point from the directrix, the position vector of this point is

$$\bar{\mathbf{r}} = \mathbf{r}_0 + \bar{u}\mathbf{d} \quad\dots\dots\dots\dots\dots\dots(25),$$

and the tangent to the line of striction is parallel to $\bar{\mathbf{r}}'$, where

$$\bar{\mathbf{r}}' = \mathbf{t} + \bar{u}\mathbf{d}' + \bar{u}'\mathbf{d}.$$

But this is perpendicular to $\mathbf{d}'$, so that

$$0 = \mathbf{d}' \cdot \bar{\mathbf{r}}' = b + \bar{u}a^2.$$

Hence
$$\bar{u} = -\frac{b}{a^2} \quad\ldots\ldots\ldots\ldots\ldots\ldots\ldots(26).$$

This determines the central point of the generator. The parametric equation of the line of striction is

$$a^2 u + b = 0.$$

Hence *if b vanishes identically the line of striction is the directrix.* This condition is that $\mathbf{d}'$ be perpendicular to $\mathbf{t}$.

Ex. Show that the line of striction cuts the generator at an angle ψ such that

$$\cot \psi = \frac{a}{D}\left\{\cos\theta - \frac{d}{ds}\left(\frac{b}{a^2}\right)\right\}.$$

70. Fundamental magnitudes. The position vector of a current point on the surface is

$$\mathbf{r} = \mathbf{r}_0 + u\mathbf{d},$$

where $\mathbf{r}_0$ and $\mathbf{d}$ are functions of the parameter s only. Hence

$$\mathbf{r}_1 = \mathbf{d},$$
$$\mathbf{r}_2 = \mathbf{t} + u\mathbf{d}',$$

so that $\qquad E = 1, \quad F = \cos\theta, \quad G = a^2 u^2 + 2bu + 1,$
$$H^2 = a^2 u^2 + 2bu + \sin^2\theta,$$

as is also evident from (21). The unit normal to the surface is

$$\mathbf{n} = \frac{\mathbf{r}_1 \times \mathbf{r}_2}{H} = \frac{1}{H}\mathbf{d} \times (\mathbf{t} + u\mathbf{d}')\ldots\ldots\ldots\ldots(27).$$

The second derivatives of $\mathbf{r}$ are

$$\mathbf{r}_{11} = 0, \quad \mathbf{r}_{12} = \mathbf{d}', \quad \mathbf{r}_{22} = \mathbf{t}' + u\mathbf{d}'',$$

so that $\qquad L = 0$

$$M = \frac{1}{H}\mathbf{d} \times (\mathbf{t} + u\mathbf{d}') \cdot \mathbf{d}' = \frac{D}{H} \left.\begin{array}{c} \\ \\ \\ \end{array}\right\} \quad\ldots\ldots\ldots\ldots(28).$$

$$N = \frac{1}{H}[\mathbf{d},\ \mathbf{t} + u\mathbf{d}',\ \mathbf{t}' + u\mathbf{d}'']$$

The specific curvature has the value

$$K = \frac{LN - M^2}{H^2} = -\frac{D^2}{H^4} \quad\ldots\ldots\ldots\ldots(29),$$

so that, *for a developable surface, K vanishes identically* (Art. 33).
Further, since K is negative, there are no elliptic points on a real
ruled surface; and, since we may write

$$H^2 = \sin^2\theta + a^2(u - \bar{u})^2 - a^2\bar{u}^2,$$

it follows that, along any generator, H^2 is least at the central
point. Therefore, *on any one generator, the second curvature is
greatest in absolute value at the central point; and, at points equidistant from this, it has equal values.*

The first curvature is given by

$$J = \frac{EN + GL - 2FM}{H^2},$$

so that $J = \frac{1}{H^3}\{[\mathbf{d},\ \mathbf{t} + u\mathbf{d}',\ \mathbf{t}' + u\mathbf{d}''] - 2D\cos\theta\}.$

71. Tangent plane. The tangent plane to a developable
surface is the same at all points of a given generator. But this
is not the case with a skew surface. We shall now show that, as
the point of contact moves along a generator from one end to the
other, the tangent plane turns through an angle of 180°. To do
this we find the inclination of the tangent plane at any point P to
the tangent plane at the central point P_0 of the same generator.
The tangent plane at the central point is called the *central plane*
of the generator.

We lose no generality by taking the *line of striction as directrix*.
Then $b = 0$, and the central point of the generator is given by
$u = 0$. We thus have

$$H^2 = a^2 u^2 + \sin^2\theta; \quad D = \pm a\sin\theta,$$

so that, for the central point,

$$H_0 = \sin\theta.$$

Similarly the unit normal at the central point is

$$\mathbf{n}_0 = \frac{\mathbf{d} \times \mathbf{t}}{\sin\theta}.$$

Let ϕ be the angle of rotation (in the sense which is positive for
the direction $\mathbf{d}$) from the central plane to the tangent plane at the
point u. This is equal to the angle of rotation from the normal $\mathbf{n}_t$

to the normal **n**, and is given by

$$\mathbf{d} \sin \phi = \mathbf{n}_0 \times \mathbf{n} = \frac{1}{H \sin \theta} (\mathbf{d} \times \mathbf{t}) \times (\mathbf{d} \times \mathbf{t} + u\mathbf{d} \times \mathbf{d}')$$

$$= \frac{u}{H \sin \theta} \{[\mathbf{d}, \mathbf{d}, \mathbf{d}']\mathbf{t} - [\mathbf{t}, \mathbf{d}, \mathbf{d}']\mathbf{d}\}$$

$$= \frac{uD}{H \sin \theta} \mathbf{d}.$$

Therefore
$$\sin \phi = \frac{uD}{H \sin \theta} = \pm \frac{au}{H} \quad \dots\dots\dots\dots(30),$$

the positive or negative sign being taken according as the surface is right-handed or left-handed. Hence, as u varies from $-\infty$ to $+\infty$, $\sin \phi$ varies continuously from -1 to $+1$, or from $+1$ to -1. Thus, *as the point of contact moves from one end of the generator to the other, the tangent plane turns through half a revolution; and the tangent planes at the ends of the generator are perpendicular to the central plane.*

Consequently $\cos \phi$ is positive, and in virtue of (30)

$$\cos \phi = \frac{1}{H} \sqrt{H^2 - a^2 u^2} = \frac{\sin \theta}{H} = \frac{|D|}{aH} \quad \dots\dots\dots(30'),$$

and therefore
$$\tan \phi = \frac{\sin \phi}{\cos \phi} = \pm \frac{au}{\sin \theta} = \frac{a^2 u}{D}$$

$$= \frac{u}{\beta} \quad \dots\dots\dots\dots\dots\dots\dots\dots(31),$$

where β is the parameter of distribution (Art. 68). Thus $\tan \phi$ is proportional to the distance of the point of contact from the central point. And, in virtue of (31), the tangent planes at two points u, U on the same generator will be perpendicular provided

$$uU = -\beta^2.$$

Thus any plane through a generator is a tangent plane at some point of the generator, and a normal plane at some other point of it. Also the points of contact of perpendicular tangent planes along a generator form an involution, with the central point as centre, and imaginary double points.

Ex. 1. *Surface of binormals.* Consider the surface generated by the binormals of a twisted curve. Take the curve itself as directrix, and let **t**, **n**, **b** be its unit tangent, principal normal and binormal respectively. Then
$\theta = \frac{\pi}{2}$, $\mathbf{d} = \mathbf{b}$, $\mathbf{d}' = -\tau\mathbf{n}$, so that

$$D = [\mathbf{t}, \ -\tau\mathbf{n}, \ \mathbf{b}] = -\tau.$$

Hence the surface is left-handed or right-handed according as the torsion of the curve is positive or negative. Further

$$a^2 = \mathbf{b}'^2 = \tau^2,$$

and $\qquad\qquad\qquad b = \mathbf{t} \cdot \mathbf{b}' = 0,$

so that the curve itself is the line of striction on the surface. The parameter of distribution is

$$\beta = \frac{D}{a^2} = -\frac{1}{\tau} = -\sigma,$$

where σ is the radius of torsion of the curve. This makes β positive when the torsion is negative.

Further $\qquad\qquad E = 1, \quad F = 0, \quad G = 1 + u^2\tau^2 = H^2,$

and the specific curvature is

$$K = -\frac{D^2}{H^4} = -\frac{\tau^2}{(1 + u^2\tau^2)^2}.$$

At a point on the curve itself the specific curvature has the value $-\tau^2$.

Ex. 2. *Surface of principal normals.* Consider the skew surface generated by the principal normals to a twisted curve. Again $\theta = \frac{\pi}{2}$, while $\mathbf{d} = \mathbf{n}$, $\mathbf{d}' = \tau\mathbf{b} - \kappa\mathbf{t}$, so that

$$D = [\mathbf{t}, \ \tau\mathbf{b} - \kappa\mathbf{t}, \ \mathbf{n}] = -\tau,$$

and the surface is therefore left-handed where τ is positive. Further,

$$a^2 = \mathbf{n}'^2 = \kappa^2 + \tau^2,$$

and $\qquad\qquad\qquad b = \mathbf{t} \cdot \mathbf{n}' = -\kappa,$

so that the distance of the central point from the curve is

$$\bar{u} = -\frac{b}{a^2} = \frac{\kappa}{\kappa^2 + \tau^2}.$$

The parameter of distribution is

$$\beta = \frac{D}{a^2} = -\frac{\tau}{\kappa^2 + \tau^2}.$$

The first order magnitudes are

$$E = 1, \quad F = 0, \quad G = (1 - u\kappa)^2 + u^2\tau^2 = H^2,$$

and the specific curvature is

$$K = -\frac{D^2}{H^4} = -\frac{\tau^2}{\{(1 - u\kappa)^2 + u^2\tau^2\}^2}.$$

At a point on the curve itself the specific curvature is $-\tau^2$.

72. Bonnet's theorem. The geodesic curvature of any curve on a surface is equal to [**n**, **r′**, **r″**]; and therefore the geodesic curvature of the directrix curve on a ruled surface is given by [**n**, **t**, **t′**] if we put $u = 0$. Now for points on the directrix

$$\mathbf{n} = \frac{\mathbf{d} \times \mathbf{t}}{H},$$

and $H = \sin \theta.$

Hence the geodesic curvature of the directrix is

$$\kappa_g = \frac{1}{\sin \theta}(\mathbf{d} \times \mathbf{t}) \times \mathbf{t} \cdot \mathbf{t′} = \frac{1}{\sin \theta}(\mathbf{d} \cdot \mathbf{t}\mathbf{t} - \mathbf{t}^2\mathbf{d}) \cdot \mathbf{t′}.$$

But the first term vanishes because $\mathbf{t} \cdot \mathbf{t′}$ is zero. Thus

$$\kappa_g = -\frac{\mathbf{d} \cdot \mathbf{t′}}{\sin \theta} = -\frac{1}{\sin \theta}\left\{ \frac{d}{ds}(\mathbf{d} \cdot \mathbf{t}) - \mathbf{d′} \cdot \mathbf{t} \right\}$$

$$= -\frac{1}{\sin \theta}\left\{ \frac{d}{ds}(\cos \theta) - b \right\}.$$

Hence the formula $\kappa_g = \dfrac{d\theta}{ds} + \dfrac{b}{\sin \theta}$ ....(32).

Now if the first member vanishes the directrix is a geodesic. If $d\theta/ds$ is zero it cuts the generators at a constant angle. If b is identically zero the directrix is the line of striction. Hence since the directrix may be chosen at pleasure, subject to the condition that it cuts all the generators, we have the following theorem, due to Bonnet:

If a curve is drawn on a ruled surface so as to intersect all the generators, then, provided it has two of the following properties, it will also have the third: (a) it is a geodesic, (b) it is the line of striction, (c) it cuts the generators at a constant angle.

Let an orthogonal trajectory of the generators be chosen as directrix. Then θ has the constant value $\dfrac{\pi}{2}$. The geodesic curvature of the directrix is then equal to b, and this vanishes where the directrix crosses the line of striction. Thus *the line of striction is the locus of the points at which the geodesic curvature of the orthogonal trajectories of the generators vanishes.*

Ex. Show that a twisted curve is a geodesic on the surface generated by its binormals.

73. Asymptotic lines. Since, for a ruled surface, $L = 0$, the differential equation of the asymptotic lines is

$$ds\,(2M\,du + N\,ds) = 0.$$

Thus the parametric curves $s = \text{const.}$, that is to say the generators, are one system of asymptotic lines. The other system (which may be referred to as the system of curved asymptotic lines) is given by

$$\frac{du}{ds} = -\frac{N}{2M} = -\frac{1}{2D}\,[\mathbf{d},\ \mathbf{t} + u\mathbf{d}',\ \mathbf{t}' + u\mathbf{d}''].$$

This equation is of the Riccati type

$$\frac{du}{ds} = Pu^2 + Qu + R,$$

in which P, Q, R are functions of s only. Its primitive is of the form *

$$u = \frac{cW + X}{cY + Z} \quad\ldots\ldots\ldots\ldots\ldots\ldots\ldots(33),$$

where c is an arbitrary constant, and W, X, Y, Z are known functions of s. This equation then gives the curved asymptotic lines, c having a different value for each member of the family.

Consider the intersections of four particular asymptotic lines c_1, c_2, c_3, c_4 with a given generator $s = \text{const.}$ Let u_1, u_2, u_3, u_4 be the points of intersection. Then by (33)

$$\frac{(u_1 - u_2)(u_3 - u_4)}{(u_1 - u_3)(u_2 - u_4)} = \frac{(c_1 - c_2)(c_3 - c_4)}{(c_1 - c_3)(c_2 - c_4)},$$

which is independent of s, and is therefore the same for all generators. Since then u is the distance measured along the generator from the directrix, this relation shows that *the cross-ratio of the four points, in which a generator is cut by four given curved asymptotic lines, is the same for all generators.*

EXAMPLES X

1. Show that the product of the specific curvatures of a ruled surface at two points on the same generator is equal to $\frac{1}{l^4}\sin^4 a$, where l is the distance between the points, and a the inclination of the tangent planes thereat.

With the notation of Art. 71, if suffixes 1 and 2 be used to distinguish the two points, we have

$$l = u_1 - u_2 = \beta\,(\tan\phi_1 - \tan\phi_2) = \frac{\beta\sin a}{\cos\phi_1\cos\phi_2}.$$

* Forsyth, *Differential Equations*, Art. 110 (3rd ed.).

Therefore $\left(\dfrac{\sin a}{l}\right)^4 = \left(\dfrac{\cos \phi_1 \cos \phi_2}{\beta}\right)^4 = \dfrac{a^8}{D^4}\left(\dfrac{D}{aH_1}\right)^4\left(\dfrac{D}{aH_2}\right)^4$

by (30') Art. 71, $\quad = \dfrac{D^4}{H_1^4 H_2^4} = K_1 K_2$

in virtue of (29).

2. Show that the normals to a ruled surface along a given generator constitute a hyperbolic paraboloid with vertex at the central point of the generator.

3. Determine the condition that the directrix be a geodesic.

4. The cross-ratio of four tangent planes to a skew surface at points of a generator is equal to the cross-ratio of the points.

5. Prove that, if the specific curvature of a ruled surface is constant, the surface is a developable.

6. Determine the condition that the line of striction may be an asymptotic line.

7. Deduce formula (32) from the value $-HnG^{-\frac{3}{2}}$ found in Art. 54 for the geodesic curvature of the parametric curve $u = \text{const.}$

8. Deduce formula (32) of Art. 72 from formula (22) of Art. 54.

9. The surface generated by the tangents to a twisted curve is a developable surface with first curvature $\tau/(u\kappa)$. Find the lines of curvature.

10. A straight line cuts a twisted curve at a constant angle and lies in the rectifying plane. Show that, on the surface which it generates, the given curve is the line of striction. Find the parameter of distribution and the specific curvature.

If $\mathbf{t}, \mathbf{n}, \mathbf{b}$ are the tangent, principal normal and binormal to the curve, we may write

$$\mathbf{d} = \frac{\mathbf{t}+c\mathbf{b}}{\sqrt{1+c^2}}, \quad \mathbf{d}' = \frac{\kappa-c\tau}{\sqrt{1+c^2}}\,\mathbf{n},$$

where c is constant. Hence $\quad b = \mathbf{d}' \bullet \mathbf{t} = 0,$
so that the curve is the line of striction. Also

$$a^2 = \mathbf{d}'^2 = \frac{(\kappa-c\tau)^2}{1+c^2},$$

$$D = \left[\mathbf{t}, \frac{\kappa-c\tau}{\sqrt{1+c^2}}\,\mathbf{n}, \frac{\mathbf{t}+c\mathbf{b}}{\sqrt{1+c^2}}\right]$$

$$= \frac{c(\kappa-c\tau)}{1+c^2}.$$

Hence the parameter of distribution is

$$\beta = \frac{D}{a^2} = \frac{c}{\kappa-c\tau}.$$

The specific curvature is

$$K = -\frac{D^2}{H^4} = -\frac{c^2(\kappa-c\tau)^2}{\{(\kappa-c\tau)^2 u^2 + c^2\}^2}.$$

From Bonnet's theorem (Art. 72) it follows that the given curve is a geodesic on the surface.

11. The *right helicoid* or *right conoid* is the surface generated by a straight line which intersects a given straight line (the axis) at right angles, and rotates about this axis with an angular velocity proportional to the velocity of the point of intersection along the axis.

If we choose the fixed axis both as directrix and as z-axis, we may write

$$x = u \cos \frac{s}{c}, \quad y = u \sin \frac{s}{c}, \quad z = s.$$

The axis is the common perpendicular to consecutive generators, and is therefore the line of striction. Hence, by Bonnet's theorem, it is also a geodesic on the surface. Moreover, with the usual notation,

$$\mathbf{t} = (0, 0, 1),$$

$$\mathbf{d} = \left(\cos \frac{s}{c}, \sin \frac{s}{c}, 0 \right), \quad \mathbf{d}' = \frac{1}{c} \left(-\sin \frac{s}{c}, \cos \frac{s}{c}, 0 \right),$$

so that

$$a^2 = \mathbf{d}'^2 = \frac{1}{c^2},$$

and

$$b = \mathbf{d}' \cdot \mathbf{t} = 0,$$

showing that the directrix is the line of striction. Similarly

$$D = [\mathbf{t}, \mathbf{d}', \mathbf{d}] = -\frac{1}{c},$$

and the parameter of distribution is

$$\beta = \frac{D}{a^2} = -c.$$

The fundamental magnitudes of the first order are

$$E = 1, \quad F = 0, \quad G = \frac{u^2 + c^2}{c^2} = H^2.$$

The specific curvature is $K = -\dfrac{D^2}{H^4} = -\dfrac{c^2}{(u^2 + c^2)^2}.$

The second order magnitudes are

$$L = 0, \quad M = \frac{D}{H} = -\frac{1}{\sqrt{u^2 + c^2}}, \quad N = 0.$$

Hence the first curvature is zero and the surface is a *minimal surface*. The principal curvatures are $\pm c/(u^2 + c^2)$. The asymptotic lines are given by

$$2M \, du \, ds = 0.$$

Hence the asymptotic lines are the generators and the curves $u = $ const.

Find also the lines of curvature.

12. If two skew surfaces have a common generator and touch at three points along it, they will touch at every point of it; also the central point and the parameter of distribution of the generator are the same for both surfaces.

13. If two skew surfaces have a common generator, and their tangent planes at three points of it are inclined at the same angle, they will be inclined at this angle at every point of the common generator.

14. The normals to a surface at points of an asymptotic line generate a skew surface whose line of striction is the asymptotic line; and the two surfaces have the same specific curvature at any point of the line.

15. Find the parameter of distribution of a generator of the cylindroid

$$z\,(x^2+y^2)=2mxy.$$

16. On the skew surface generated by the line

$$x+yt=3t\,(1+t^2),\quad y+2zt=t^2\,(3+4t^2),$$

prove that the parameter of distribution of a generator is $\frac{3}{2}\,(1+2t^2)^2$, and that the line of striction is the curve

$$x=3t,\quad y=3t^2,\quad z=2t^3.$$

17. The line of striction on an hyperboloid of revolution of one sheet is the principal circular section.

18. The right helicoid is the only ruled surface whose generators are the principal normals of their orthogonal trajectories.

19. If two of the curved asymptotic lines of a skew surface are orthogonal trajectories of the generators, they are Bertrand curves; if all of them are orthogonal trajectories, the surface is a right helicoid.

20. The right helicoid is the only ruled surface each of whose lines of curvature cuts the generators at a constant angle. On any other skew surface there are in general four lines of curvature which have this property.

21. The line of striction of a skew surface is an orthogonal trajectory of the generators only if the latter are the binormals of a curve, or if the surface is a right conoid.

22. If the lines of curvature of one family on a ruled surface are such that the segments of the generators between two of them are of the same length, the parameter of distribution is constant, and the line of striction is a line of curvature.

NOTE. The author has recently shown that a family of curves on any surface possesses a line of striction, and that the theorem of Art. 72 is true for a family of geodesics on any surface. See Art. 126 below.

The remaining chapters of the book may be read in any order.

CHAPTER VIII

EVOLUTE OR SURFACE OF CENTRES.
PARALLEL SURFACES

SURFACE OF CENTRES

74. Centro-surface. We have already seen (Art. 29) that consecutive normals along a line of curvature intersect, the point of intersection being the corresponding centre of curvature. The locus of the centres of curvature for all points of a given surface S is called the *surface of centres* or *centro-surface* of S. In general it consists of two sheets, corresponding to the two families of lines of curvature.

Along any one line of curvature, C, the normals to the surface generate a developable surface whose edge of regression is the locus of the centres of curvature along C. All these normals touch the edge of regression, which is therefore an evolute of C. If now we consider all members of that family of lines of curvature to which C belongs, the locus of their edges of regression is a surface, which is one sheet of the surface of centres. Similarly from the other family of lines of curvature we have another family of edges of regression which lie on the second sheet of the centro-surface.

Let PQ, RT be consecutive lines of curvature of the first system, and PR, QT consecutive lines of curvature of the second system. The normals

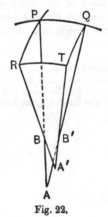

Fig. 22.

to the surface at P and Q intersect at a point A on the *first* sheet of the centro-surface, and those at R and T intersect at another point A' on the same sheet. Similarly the normals at P, R intersect at a point B, and those at Q, T intersect at another B', both on the *second* sheet. Thus PA, PB are equal in magnitude to the principal radii of curvature at P. The normal PA is a tangent to the edge of regression corresponding to the curve PQ, and is therefore a tangent to the first sheet of the centro-surface at A. Similarly RA' is a tangent to the same sheet at A'. And as A, A' are consecutive points to this sheet, AA' is also a tangent line to the sheet. But these three tangents are all in the plane PRB. Hence in the limit, as R tends to coincidence with P, the normal plane at P in the direction PR is tangential to the first sheet of the centro-surface at A. Thus *the normal at A to the first sheet of the surface of centres is parallel to the tangent at P to the corresponding line of curvature PQ.* Similarly the normal at B to the second sheet of the centro-surface is parallel to the tangent at P to the other line of curvature PR. The normals to the two sheets of the centro-surface at corresponding points A, B are therefore perpendicular to each other.

Because the surface of centres is the envelope of the principal normal planes, and is composed of evolutes of the lines of curvature on S, it is often called the *evolute* of S. These evolutes of the lines of curvature on S, which are the edges of regression of the developables generated by the normals, are also geodesics on the surface of centres. To prove this consider the edge of regression of the surface generated by the normals along the line of curvature PQ. The osculating plane of this curve at A is the plane of consecutive normals PA, QA to the surface. Hence it contains the tangent at P to the curve PQ, and therefore also the normal at A to the first sheet of the centro-surface. *The edge of regression is thus a geodesic on the centro-surface.* Similarly the edge of regression of the developable generated by the normals along the line of curvature PR is a geodesic on the second sheet of the evolute.

It is easy to show that *the orthogonal trajectories of these regressional geodesics are the curves on the first sheet which correspond to the lines $\alpha = const.$ on S, and the curves on the second sheet corresponding to the lines $\beta = const.$,* where α, β are the principal radii of curvature for the surface S. For, considering the first sheet of the evolute, let T' be one of these orthogonal trajectories, and T the corresponding curve on S. Then the normals to S along T generate a developable surface on which T and T' are orthogonal trajectories of the generators, and therefore intercept between them segments of the generators of constant length (Art. 56). Thus along the curve T on S the radius of curvature a is constant.

Moreover, *the curves on either sheet of the evolute which correspond to the lines of curvature on S, form a conjugate system.* For convenience of description let the lines of curvature be referred to as parametric curves, PQ belonging to the system $v=$ const., and PR to the system $u=$ const. The normal at P touches both sheets of the evolute. As a member of the family of normals along PR it is a tangent to a regressional geodesic $u=$ const. on the second sheet. As a member of the family of normals along PQ it is a tangent to a regressional geodesic $v=$ const. on the first sheet, and touches the second sheet at a point on the corresponding line $v=$ const. Thus the normals along PQ form a developable surface, whose generators touch the second sheet along a line $v=$ const., and are tangents there to the lines $u=$ const. Hence (Art. 35) the parametric curves are conjugate on the second sheet; and these are the curves corresponding to the lines of curvature on S. Similarly the theorem may be proved for the first sheet.

All these properties will be proved analytically in the following Art. Meanwhile we may observe in passing that it follows from the last theorem and Beltrami's theorem (Art. 56 Ex.) that the centres of geodesic curvature of the orthogonal trajectories of the regressional geodesics on either sheet of the evolute are the corresponding points on the other sheet. For, on the second sheet, the lines $v=$ const. are conjugate to the geodesics $u=$ const. And the tangents to these geodesics along a line $v=$ const. form a developable whose edge of regression is a geodesic on the first sheet. But, by Beltrami's theorem, each point of this edge of regression is the centre of geodesic curvature of the orthogonal trajectory of the geodesics $u=$ const. at the corresponding point on the second sheet: hence the result. It follows that the radius of geodesic curvature of the orthogonal trajectory is numerically equal to the difference between the principal radii of curvature of the surface S.

75. Fundamental magnitudes. The same results may be obtained analytically as follows. Let the lines of curvature on S be taken as parametric curves. If α, β are the principal radii of curvature at the point $\mathbf{r}$ on S, the corresponding point A on the *first sheet* of the centro-surface is

$$\bar{\mathbf{r}} = \mathbf{r} + \alpha \mathbf{n} \quad\dots\dots\dots\dots\dots(1).$$

Now since A is the centre of curvature for $v = \text{const.}$ it follows that $\bar{\mathbf{r}}$ and α are constant for one differentiation with respect to u. Thus

$$\left.\begin{array}{r} \mathbf{r}_1 + \alpha \mathbf{n}_1 = 0 \\ \mathbf{r}_2 + \beta \mathbf{n}_2 = 0 \end{array}\right\} \quad\dots\dots\dots\dots\dots(2).$$

Similarly

These are the equivalent of Rodrigues' formula already proved in Art. 30. In virtue of these we obtain from (1)

$$\left.\begin{array}{l} \bar{\mathbf{r}}_1 = \alpha_1 \mathbf{n} \\[2mm] \bar{\mathbf{r}}_2 = \left(1 - \dfrac{\alpha}{\beta}\right) \mathbf{r}_2 + \alpha_2 \mathbf{n} \end{array}\right\} \quad\dots\dots\dots\dots\dots(3),$$

so that the magnitudes for the first sheet of the evolute are

$$\bar{E} = \alpha_1{}^2, \quad \bar{F} = \alpha_1 \alpha_2, \quad \bar{G} = \alpha_2{}^2 + G\left(1 - \frac{\alpha}{\beta}\right)^2, \quad \bar{H}^2 = \alpha_1{}^2 G\left(1 - \frac{\alpha}{\beta}\right)^2.$$

The square of the linear element for the first sheet is

$$d\bar{s}^2 = \bar{E}\,du^2 + 2\bar{F}\,du\,dv + \bar{G}\,dv^2$$
$$= d\alpha^2 + G\left(1 - \frac{\alpha}{\beta}\right)^2 dv^2 \quad\dots\dots\dots\dots(4),$$

which is of the geodesic form. Hence the curves $v = \text{const.}$ are geodesics on the first sheet of the centro-surface. These are the edges of regression of the developables generated by the normals along the lines of curvature $v = \text{const.}$ on S. The orthogonal trajectories of these regressional geodesics are the curves $\alpha = \text{const.}$, which agrees with the result proved in the preceding Art.

The unit normal $\bar{\mathbf{n}}$ to the first sheet is given by

$$\bar{H}\bar{\mathbf{n}} = \bar{\mathbf{r}}_1 \times \bar{\mathbf{r}}_2 = \alpha_1\left(1 - \frac{\alpha}{\beta}\right)\mathbf{n} \times \mathbf{r}_2.$$

But $\mathbf{r}_1/\sqrt{E}$, $\mathbf{r}_2/\sqrt{G}$ and $\mathbf{n}$ form a right-handed system of unit vectors. Consequently the last equation may be written

$$\bar{H}\bar{\mathbf{n}} = -\alpha_1\left(1 - \frac{\alpha}{\beta}\right)\sqrt{G}\left(\frac{\mathbf{r}_1}{\sqrt{E}}\right) \quad\dots\dots\dots(5),$$

agreeing with the result, previously established, that the normal at A to the first sheet of the evolute is parallel to the tangent at P to the line of curvature PQ. We may express this

$$\bar{\mathbf{n}} = \epsilon\mathbf{r}_1/\sqrt{E} \quad\dots\dots\dots\dots\dots\dots(6),$$

where ϵ is ± 1 according as $\alpha_1\left(1 - \dfrac{\alpha}{\beta}\right)$ is negative or positive.

The fundamental magnitudes of the second order for the first sheet of the evolute may now be calculated. For

$$\bar{L} = \bar{\mathbf{n}}\cdot\bar{\mathbf{r}}_{11} = \frac{\epsilon\mathbf{r}_1}{\sqrt{E}}\cdot(\alpha_1\mathbf{n}_1 + \alpha_{11}\mathbf{n}) = -\epsilon\alpha_1\frac{\mathbf{r}_1}{\sqrt{E}}\cdot\frac{\mathbf{r}_1}{\alpha}$$

in virtue of (2). Hence finally

$$\bar{L} = -\epsilon\sqrt{E}\alpha_1/\alpha.$$

Similarly $\bar{M} = \bar{\mathbf{n}}\cdot\bar{\mathbf{r}}_{12} = \dfrac{\epsilon\mathbf{r}_1}{\sqrt{E}}\cdot(\alpha_1\mathbf{n}_2 + \alpha_{12}\mathbf{n}) = 0$

in virtue of (2). And

$$\bar{N} = \bar{\mathbf{n}}\cdot\bar{\mathbf{r}}_{22} = \frac{\epsilon}{\sqrt{E}}\left(1 - \frac{\alpha}{\beta}\right)\mathbf{r}_1\cdot\mathbf{r}_{22} \qquad\text{by (3),}$$

all the other scalar products vanishing. Now

$$\mathbf{r}_1\cdot\mathbf{r}_{22} = \frac{\partial}{\partial v}(\mathbf{r}_1\cdot\mathbf{r}_2) - \mathbf{r}_{12}\cdot\mathbf{r}_2 = F_2 - \tfrac{1}{2}G_1 = -\tfrac{1}{2}G_1,$$

because the parametric curves are orthogonal. Also since the parametric curves are lines of curvature

$$-\tfrac{1}{2}G_1 = \frac{\alpha G\beta_1}{\beta(\beta-\alpha)}. \qquad\text{[Cf. Ex. 2 below.]}$$

On substituting this value in the formula for $\bar{N}$ we have

$$\bar{N} = \frac{\epsilon\alpha G\beta_1}{\beta^2\sqrt{E}}.$$

Collecting the results thus established we have

$$\bar{L} = -\epsilon\sqrt{E}\frac{\alpha_1}{\alpha}, \quad \bar{M} = 0, \quad \bar{N} = \frac{\epsilon\alpha G}{\sqrt{E}}\frac{\beta_1}{\beta^2}\dots\dots\dots(7).$$

Since $\bar{M} = 0$ it follows that the parametric curves on the centro-surface form a conjugate system. Thus *the curves on the evolute, which correspond to lines of curvature on the original surface, are conjugate*, but (in general) are not lines of curvature because $\bar{F}$ is not zero.

The fundamental magnitudes for the *second sheet* of the centro-surface are obtainable from the above by interchanging simultaneously u and v, E and G, L and N, α and β. Thus for the second sheet we have the first order quantities

$$\bar{E}' = \beta_1{}^2 + \left(1 - \frac{\beta}{\alpha}\right)^2 E, \quad \bar{F}' = \beta_1 \beta_2, \quad \bar{G}' = \beta_2{}^2,$$

and the second order quantities

$$\bar{L}' = \frac{\epsilon' \beta E}{\sqrt{G}} \frac{\alpha_2}{\alpha^2}, \quad \bar{M}' = 0, \quad \bar{N}' = - \epsilon' \sqrt{G} \frac{\beta_2}{\beta} \quad \ldots\ldots\ldots(8),$$

where ϵ' is equal to ± 1 according as $\beta_2 \left(1 - \dfrac{\beta}{\alpha}\right)$ is negative or positive.

The specific curvature for the first sheet of the evolute is

$$\bar{K} = \frac{\bar{L}\bar{N}}{\bar{H}^2} = - \frac{1}{(\alpha - \beta)^2} \frac{\beta_1}{\alpha_1} \quad \ldots\ldots\ldots\ldots\ldots(9),$$

and for the second sheet

$$\bar{K}' = - \frac{1}{(\alpha - \beta)^2} \frac{\alpha_2}{\beta_2} \quad \ldots\ldots\ldots\ldots\ldots\ldots(9').$$

Ex. 1. Write down the expressions for the first curvatures of the two sheets of the evolute.

Ex. 2. Prove that, if the lines of curvature are parametric curves,

$$G_1 = \frac{2aG\beta_1}{\beta(a - \beta)}, \quad E_2 = \frac{2\beta E a_2}{\alpha(\beta - \alpha)}.$$

It follows from the data that $F = M = 0$ and

$$a = \frac{E}{L}, \quad \beta = \frac{G}{N}.$$

From the Mainardi-Codazzi relation (8) of Art. 43 it then follows that

$$- \frac{\partial}{\partial u}\left(\frac{G}{\beta}\right) = nL - \mu N,$$

and therefore

$$\frac{G\beta_1}{\beta^2} - \frac{G_1}{\beta} = - \frac{GG_1}{2H^2}\frac{E}{a} - \frac{EG_1}{2H^2}\frac{G}{\beta}.$$

Then, since $H^2 = EG$, this reduces to the required formula

$$G_1 = \frac{2aG\beta_1}{\beta(a - \beta)}.$$

The other result follows in like manner from the relation (7) of Art. 43.

Ex. 3. Prove the formulae given above for the fundamental magnitudes of the second sheet of the centro-surface.

76. Weingarten surfaces. The *asymptotic lines* on the first sheet of the centro-surface are found from the equation

$$\overline{L}\,du^2 + 2\overline{M}\,du\,dv + \overline{N}\,dv^2 = 0.$$

On substitution of the values of the fundamental magnitudes found above, this reduces to

$$E\alpha_1\beta^2\,du^2 - G\beta_1\alpha^2\,dv^2 = 0 \ldots\ldots\ldots\ldots\ldots(10).$$

Similarly the asymptotic lines on the second sheet are given by

$$E\alpha_2\beta^2\,du^2 - G\beta_2\alpha^2\,dv^2 = 0 \ \ldots\ldots\ldots\ldots(10').$$

The asymptotic lines on the two sheets will therefore correspond if these two equations are identical. This will be the case if

$$\alpha_1\beta_2 = \alpha_2\beta_1,$$

that is to say, if α, β are connected by some functional relation

$$f(\alpha, \beta) = 0.$$

Surfaces with this property are called *Weingarten surfaces*. The above analysis is reversible, so that we have the theorem: *If there exists a functional relation between the principal curvatures of a surface, the asymptotic lines on the two sheets of its evolute correspond.*

Weingarten surfaces are exemplified by surfaces of constant specific curvature K, surfaces of constant first curvature J, or more generally by surfaces in which there is any functional relation $f(J, K) = 0$ between these two curvatures. Since, on a Weingarten surface, either principal radius of curvature may be regarded as a function of the other, the formulae found above for the specific curvatures of the two sheets of the centro-surface may be written

$$\overline{K} = -\frac{1}{(\alpha - \beta)^2}\frac{d\beta}{d\alpha}\Bigg\}$$

and

$$\overline{K}' = -\frac{1}{(\alpha - \beta)^2}\frac{d\alpha}{d\beta}\Bigg\} \ldots\ldots\ldots\ldots\ldots(11).$$

Thus, for any Weingarten surface,

$$\overline{K}\,\overline{K}' = \frac{1}{(\alpha - \beta)^4} \ \ldots\ldots\ldots\ldots(12).$$

Consider the particular case in which the functional relation between the principal radii of curvature is

$$\alpha - \beta = c \ldots\ldots\ldots\ldots\ldots(13),$$

where c is a constant. From this it follows that
$$d\alpha = d\beta,$$
so that the formulae (11) become
$$\overline{K} = \overline{K}' = -\frac{1}{c^2} \quad\dots\dots\dots\dots\dots\dots(14).$$

Surfaces of constant negative specific curvature are called *pseudo-spherical* surfaces. Hence *the two sheets of the evolute of a surface, whose principal radii have a constant difference, are pseudo-spherical surfaces.*

For Weingarten surfaces of the class (13), not only do the asymptotic lines on the two sheets of the centro-surface correspond, but corresponding portions are of equal length. For, on the first sheet, the square of the linear element is
$$d\bar{s}^2 = d\alpha^2 + \frac{G}{\beta^2}(\alpha - \beta)^2 \, dv^2,$$
and on the second sheet
$$d\bar{s}'^2 = d\beta^2 + \frac{E}{\alpha^2}(\alpha - \beta)^2 \, du^2.$$

But, in virtue of (10) and (10'), since $\alpha_1 = \beta_1$ and $\alpha_2 = \beta_2$, it follows that along asymptotic lines of the evolute,
$$\frac{E}{\alpha^2} \, du^2 = \frac{G}{\beta^2} \, dv^2.$$
Hence
$$d\bar{s}^2 - d\bar{s}'^2 = d\alpha^2 - d\beta^2 = 0,$$
showing that $d\bar{s} = d\bar{s}'$. Thus corresponding elements of asymptotic lines on the two sheets of the evolute are equal in length, and the theorem is proved.

If we consider the possibility of the asymptotic lines of the surface S corresponding with those of the first sheet of the evolute, we seek to identify (10) with the equation of the asymptotic lines of the surface S. Now since the lines of curvature are parametric curves on S, its asymptotic lines are given by
$$L du^2 + N dv^2 = 0,$$
that is
$$\frac{E}{\alpha} \, du^2 + \frac{G}{\beta} \, dv^2 = 0.$$

This equation will be identical with (10) provided
$$\alpha_1\beta + \alpha\beta_1 = 0,$$
that is
$$\frac{\partial}{\partial u}(\alpha\beta) = 0 \quad\dots\dots\dots\dots\dots\dots(15).$$

This requires K to be constant along the lines of curvature $v = \text{const.}$ Thus *in order that the asymptotic lines on a surface S may correspond with those on one sheet of its centro-surface, the lines of curvature on S corresponding to this sheet must be lines of constant specific curvature.* Hence, in order that the asymptotic lines on S may correspond to those on each sheet of its evolute, the specific curvature of S must be constant.

77. Lines of curvature. We have seen that the lines of curvature on a surface S do not in general correspond with those on its centro-surface. We naturally enquire if the lines of curvature on one sheet of the evolute correspond with those on the other. If in the general differential equation of the lines of curvature on a surface,

$$(EM - FL)\,du^2 + (EN - GL)\,du\,dv + (FN - GM)\,dv^2 = 0,$$

we substitute the magnitudes belonging to the *first sheet* of the centro-surface we obtain, after reduction, the differential equation of the lines of curvature on this sheet, in the form

$$E\beta^2 \alpha_1 \alpha_2 du^2 + G\alpha^2 \alpha_2 \beta_1 dv^2$$
$$+ \{E\beta^2 \alpha_2^2 + G\alpha^2 \alpha_1 \beta_1 + EG\,(\alpha - \beta)^2\}\,du\,dv = 0.$$

Similarly on using the fundamental magnitudes for the *second sheet* we find the differential equation of its lines of curvature to be

$$E\beta^2 \beta_1 \alpha_2 du^2 + G\alpha^2 \beta_1 \beta_2 dv^2$$
$$+ \{E\beta^2 \alpha_2 \beta_2 + G\alpha^2 \beta_1^2 + EG\,(\alpha - \beta)^2\}\,du\,dv = 0.$$

The lines of curvature on the two sheets will correspond if these two equations are identical. The necessary and sufficient conditions for this are

$$\alpha_1 = \beta_1 \quad \text{and} \quad \alpha_2 = \beta_2,$$

that is

$$\frac{\partial \alpha}{\partial u} = \frac{\partial \beta}{\partial u} \quad \text{and} \quad \frac{\partial \alpha}{\partial v} = \frac{\partial \beta}{\partial v},$$

whence

$$\alpha - \beta = c,$$

where c is constant. Hence *only in the case of the Weingarten surfaces, on which the principal radii differ by a constant, do the lines of curvature on the two sheets of the centro-surface correspond.* This theorem is due to Ribaucour.

78. Degenerate evolute. In particular instances either sheet of the evolute may degenerate into a curve. In such a case the edge of regression of the developable generated by the normals along a line of curvature becomes a single point of that curve. We proceed to enquire under what conditions the normals to a surface S will all intersect a given curve C.

Let $\mathbf{r}$ be a point on the surface S, $\mathbf{n}$ the unit normal there, and $\bar{\mathbf{r}}$ the point in which this normal cuts the curve C. Then we may write

$$\bar{\mathbf{r}} = \mathbf{r} + t\mathbf{n}$$

or $\qquad\qquad \mathbf{r} = \bar{\mathbf{r}} - t\mathbf{n} \qquad\dots\dots\dots\dots\dots\dots\dots(16).$

Let the arc-length u of C be chosen as one of the parameters. Then $\bar{\mathbf{r}}$ is a function of u only; but the other quantities are functions of u and another parameter v. Now the normal $\mathbf{n}$ to the surface S is perpendicular to both $\mathbf{r}_1$ and $\mathbf{r}_2$. It follows then from (16) that

$$\mathbf{n} \cdot (\bar{\mathbf{r}}_1 - t_1\mathbf{n} - t\mathbf{n}_1) = 0$$

and $\qquad\qquad \mathbf{n} \cdot (t_2\mathbf{n} + t\mathbf{n}_2) = 0,$

which are equivalent to

$$t_2 = 0, \quad t_1 = \mathbf{n} \cdot \bar{\mathbf{r}}_1 = \cos\theta, \qquad\dots\dots\dots\dots\dots(17),$$

where θ is the inclination of the normal to the tangent to the curve C. Since then

$$\frac{\partial}{\partial v}(\cos\theta) = \frac{\partial t_1}{\partial v} = \frac{\partial t_2}{\partial u} = 0,$$

it follows that $\cos\theta$ is a function of u only. Thus the normals to S, which meet at a point of the curve C, form a right circular cone whose semi-vertical angle θ changes as the point moves along the curve. These intersecting normals emanate from a line of curvature on S, which must then be circular. Thus *the surface S has a system of circular lines of curvature.* And, further, the sphere described with centre at the point of concurrence of the normals, and passing through the feet of these normals, will touch S along one of the circular lines of curvature. Thus S *is the envelope of a singly infinite family of spheres with centres on the curve C.*

Conversely, if a surface S has a system of circular lines of curvature, the normals along one of these generate a circular cone, whose vertex lies on a curve C to which the corresponding sheet of the evolute degenerates. The surface S is then the envelope of a singly infinite family of spheres with centres on C.

If both systems of lines of curvature of S are circular, each sheet of the evolute degenerates to a curve. Then, from the preceding argument, it follows that each of these curves lies on a singly infinite family of circular cones whose axes are tangents to the other curve. Surfaces of this nature are called *Dupin's Cyclides*.

PARALLEL SURFACES

79. Parallel surfaces. A surface, which is at a constant distance along the normal from another surface S, is said to be *parallel* to S. As the constant distance may be chosen arbitrarily, the number of such parallel surfaces is infinite. If $\mathbf{r}$ is the current point on the surface S, $\mathbf{n}$ the unit normal to that surface, and c the constant distance, the corresponding point on the parallel surface is

$$\bar{\mathbf{r}} = \mathbf{r} + c\mathbf{n} \dots\dots\dots\dots\dots\dots(18).$$

Let the lines of curvature on S be taken as parametric curves, so that $F = 0$ and $M = 0$. Then if α, β are the principal radii of curvature on S we have, in virtue of (2),

$$\bar{\mathbf{r}}_1 = \mathbf{r}_1 + c\mathbf{n}_1 = -\frac{(c-\alpha)}{\alpha}\mathbf{r}_1$$

and
$$\bar{\mathbf{r}}_2 = \mathbf{r}_2 + c\mathbf{n}_2 = -\frac{(c-\beta)}{\beta}\mathbf{r}_2 \quad\Bigg\} \dots\dots\dots\dots(19).$$

The magnitudes of the first order for the parallel surface are therefore

$$\bar{E} = \left(\frac{c-\alpha}{\alpha}\right)^2 E, \quad \bar{F} = 0, \quad \bar{G} = \left(\frac{c-\beta}{\beta}\right)^2 G \dots\dots\dots(20),$$

and
$$\bar{H}^2 = \frac{(c-\alpha)^2(c-\beta)^2}{\alpha^2\beta^2} EG.$$

The unit normal to the parallel surface is given by

$$\bar{H}\bar{\mathbf{n}} = \bar{\mathbf{r}}_1 \times \bar{\mathbf{r}}_2 = \frac{(c-\alpha)(c-\beta)}{\alpha\beta} H\mathbf{n}.$$

Thus *the normals to the two surfaces at corresponding points are parallel*; and we may write

$$\bar{\mathbf{n}} = \epsilon\mathbf{n},$$

where ϵ is equal to ± 1 according as $\dfrac{(c-\alpha)(c-\beta)}{\alpha\beta}$ is positive or negative.

For the magnitudes of the second order on the parallel surface we have

$$\bar{L} = \epsilon \mathbf{n} \cdot \left\{ -\frac{(c-\alpha)}{\alpha}\, \mathbf{r}_{11} - \mathbf{r}_1 \frac{\partial}{\partial u}\left(\frac{c-\alpha}{\alpha}\right)\right\} = -\epsilon\,\frac{(c-\alpha)}{\alpha}\,L = -\epsilon\,\frac{(c-\alpha)}{\alpha^2}\,E.$$

Similarly

$$\bar{M} = \epsilon \mathbf{n} \cdot \left\{ -\frac{(c-\alpha)}{\alpha}\, \mathbf{r}_{12} - \mathbf{r}_1 \frac{\partial}{\partial u}\left(\frac{c-\alpha}{\alpha}\right)\right\} = -\epsilon\,\frac{(c-\alpha)}{\alpha}\,M = 0,$$

and

$$\bar{N} = \epsilon \mathbf{n} \cdot \left\{ -\frac{(c-\beta)}{\beta}\, \mathbf{r}_{22} - \mathbf{r}_2 \frac{\partial}{\partial v}\left(\frac{c-\beta}{\beta}\right)\right\} = -\epsilon\,\frac{(c-\beta)}{\beta}\,N = -\epsilon\,\frac{(c-\beta)}{\beta^2}\,G.$$

Thus $\bar{F} = \bar{M} = 0$, so that the parametric curves are lines of curvature on the parallel surface also. Hence *the lines of curvature on the parallel surface correspond to those on the original surface, and their tangents at corresponding points are parallel*, since $\bar{\mathbf{r}}_1$ is parallel to $\mathbf{r}_1$, and $\bar{\mathbf{r}}_2$ is parallel to $\mathbf{r}_2$.

80. Curvature. The principal radii of curvature for the parallel surface are

$$\left.\begin{aligned}\bar{\alpha} &= \bar{E}/\bar{L} = \epsilon\,(\alpha - c)\\ \bar{\beta} &= \bar{G}/\bar{N} = \epsilon\,(\beta - c)\end{aligned}\right\} \quad \cdots\cdots\cdots\cdots\cdots(21),$$

and

as we should expect. The first curvature is therefore

$$\bar{J} = \frac{1}{\epsilon}\left(\frac{1}{\alpha - c} + \frac{1}{\beta - c}\right) = \frac{\epsilon\,(J - 2cK)}{1 - cJ + c^2 K},$$

and the second curvature

$$\bar{K} = \frac{1}{\epsilon^2\,(\alpha - c)\,(\beta - c)} = \frac{K}{1 - cJ + c^2 K}.$$

If the specific curvature K of the original surface is constant and equal to $\dfrac{1}{a^2}$, and we take $c = \pm\,a$, we have

$$\bar{J} = \mp\,\epsilon\,\frac{1}{a}.$$

Thus *with every surface of constant second curvature $\dfrac{1}{a^2}$ there are associated two surfaces of constant first curvature $\pm\dfrac{1}{a}$, which are parallel to the former and distant $\pm\,a$ from it.* This theorem is due to Bonnet.

Similarly if J is constant and equal to $\dfrac{1}{a}$, and we take $c = a$, we find

$$\overline{K} = \frac{1}{a^2} = \text{const.}$$

Thus *with a surface of constant first curvature* $\dfrac{1}{a}$ *there is associated a parallel surface of constant second curvature* $\dfrac{1}{a^2}$ *at a distance a from it.*

The asymptotic lines on the parallel surface are given by

$$\overline{L}du^2 + \overline{N}dv^2 = 0,$$

which reduces to

$$\beta^2 (c - \alpha) Edu^2 + \alpha^2 (c - \beta) Gdv^2 = 0 \quad \ldots\ldots\ldots(22).$$

Hence they do not correspond with the asymptotic lines on the original surface, which are given by

$$Ldu^2 + Ndv^2 = 0$$

or

$$\beta Edu^2 + \alpha Gdv^2 = 0.$$

81. Involutes of a surface. We have seen that the normals to a surface are tangents to a family of geodesics on each sheet of the centro-surface. We now proceed to show that *the tangents to a singly infinite family of geodesics on a given surface are normals to a family of parallel surfaces.*

Let the family of geodesics be taken as the curves $v = \text{const.}$ and their orthogonal trajectories as the curves $u = \text{const.}$ Then we may choose u so that the square of the linear element has the geodesic form

$$ds^2 = du^2 + Gdv^2.$$

An involute of a geodesic $v = \text{const.}$ is the locus of a point whose position vector $\overline{r}$ is given by

$$\overline{r} = r + (c - u)\, r_1 \quad \ldots\ldots\ldots\ldots\ldots\ldots(23),$$

where c is constant, and r a point on the geodesic. We shall prove that, for a given value of c, the locus of these involutes is a surface $\overline{S}$ cutting orthogonally all the tangents to the family of geodesics.

From (23) it follows that

$$\overline{r}_1 = (c - u)\, r_{11},$$
$$\overline{r}_2 = r_2 + (c - u)\, r_{12}.$$

Using the values of l, m, λ, etc. found in Art. 56 when ds^2 has the geodesic form, we may write this

$$\left. \begin{aligned} \bar{\mathbf{r}}_1 &= (c-u)\,L\mathbf{n} \\ \bar{\mathbf{r}}_2 &= \mathbf{r}_2 + (c-u)\left(M\mathbf{n} + \frac{G_1}{2G}\mathbf{r}_2\right) \end{aligned} \right\} \quad \dots\dots\dots\dots(24).$$

Hence the unit normal $\bar{\mathbf{n}}$ to the locus of the involutes is given by

$$\bar{H}\bar{\mathbf{n}} = \bar{\mathbf{r}}_1 \times \bar{\mathbf{r}}_2 = (c-u)\,L\mathbf{n} \times \mathbf{r}_2 + (c-u)^2\frac{LG_1}{2G}\mathbf{n} \times \mathbf{r}_2,$$

and is therefore parallel to $\mathbf{r}_1$. Thus the surface $\bar{S}$ is normal to the tangents to the family of geodesics on the given surface S. It is called an *involute* of S with respect to this family of geodesics. And, since the value of the constant c may be chosen arbitrarily, the involutes are infinite in number and constitute a family of parallel surfaces.

With respect to any one of these involutes $\bar{S}$, the original surface S forms one sheet of the evolute. The family of geodesics on S are the edges of regression of the developables generated by the normals along one family of lines of curvature on $\bar{S}$. The orthogonal trajectories of the geodesics correspond to the lines on $\bar{S}$ along which one of the principal radii of curvature is constant. The second sheet of the evolute of $\bar{S}$ is (Art. 74) the locus of the centres of geodesic curvature of these orthogonal trajectories of the given family of geodesics on S. This second sheet is called the *complementary surface* to S with respect to that family of geodesics. From the proof of Beltrami's theorem (Art. 56, Ex.) it follows that, with the above choice of parametric lines on S, the position vector of the point $\mathbf{R}$ on the complementary surface corresponding to the point $\mathbf{r}$ on S is given by

$$\mathbf{R} = \mathbf{r} - \frac{2G}{G_1}\mathbf{r}_1 \dots\dots\dots\dots\dots\dots(25).$$

Ex. 1. Calculate the fundamental magnitudes for an involute of a given surface.

Ex. 2. Prove from (25) that the normal to the complementary surface is parallel to $\mathbf{r}_2$.

Ex. 3. Show that surfaces parallel to a surface of revolution are surfaces of revolution.

Ex. 4. Show that null lines on two parallel surfaces do not (in general) correspond.

82. Inverse surface. Consider next the surface $\bar{S}$ which is derived from a given surface S by inversion. Let the centre of inversion be taken as origin. Then, if c is the radius of inversion, the position vector $\bar{\mathbf{r}}$ of a point on the inverse surface, corresponding to the point $\mathbf{r}$ on S, has the direction of $\mathbf{r}$ and the magnitude c^2/r. It is therefore given by

$$\bar{\mathbf{r}} = \frac{c^2}{r^2}\,\mathbf{r} \quad\ldots\ldots\ldots\ldots\ldots\ldots(26).$$

Hence
$$\left.\begin{aligned}
\bar{\mathbf{r}}_1 &= \frac{c^2}{r^2}\mathbf{r}_1 - \frac{2c^2}{r^3}\,r_1\mathbf{r}\\[2mm]
\bar{\mathbf{r}}_2 &= \frac{c^2}{r^2}\mathbf{r}_2 - \frac{2c^2}{r^3}\,r_2\mathbf{r}
\end{aligned}\right\}\quad\ldots\ldots\ldots\ldots\ldots(27).$$

Also by differentiating the identity $\mathbf{r}^2 = r^2$ with respect to the parameters we have

$$\mathbf{r}\cdot\mathbf{r}_1 = rr_1, \quad \mathbf{r}\cdot\mathbf{r}_2 = rr_2 \quad\ldots\ldots\ldots\ldots\ldots(28).$$

The first order magnitudes for the inverse surface are obtained by squaring and multiplying (27). Then, in virtue of (28), we find

$$\left.\begin{aligned}
\bar{E} &= \bar{\mathbf{r}}_1{}^2 = \frac{c^4}{r^4}\,E\\[2mm]
\bar{F} &= \bar{\mathbf{r}}_1\cdot\bar{\mathbf{r}}_2 = \frac{c^4}{r^4}\,F\\[2mm]
\bar{G} &= \bar{\mathbf{r}}_2{}^2 = \frac{c^4}{r^4}\,G
\end{aligned}\right\}\quad\ldots\ldots\ldots\ldots\ldots(29),$$

and therefore
$$\bar{H} = \frac{c^4}{r^4}\,H.$$

Since the first order magnitudes for $\bar{S}$ are proportional to those for S it follows that *the angle between any two curves is unaltered by inversion*, and also that null lines are inverted into null lines.

The unit normal to the inverse surface is found from the formula

$$\bar{H}\bar{\mathbf{n}} = \bar{\mathbf{r}}_1 \times \bar{\mathbf{r}}_2 = \frac{c^4}{r^4}H\mathbf{n} - \frac{2c^4}{r^5}\,(r_2\mathbf{r}_1 \times \mathbf{r} - r_1\mathbf{r}_2 \times \mathbf{r}).$$

Now by (28) the expression in brackets is equal to

$$\frac{1}{r}(\mathbf{r} \cdot \mathbf{r}_2 \mathbf{r}_1 - \mathbf{r} \cdot \mathbf{r}_1 \mathbf{r}_2) \times \mathbf{r} = \frac{1}{r}\{\mathbf{r} \times (\mathbf{r}_1 \times \mathbf{r}_2)\} \times \mathbf{r}$$

$$= \frac{H}{r}(\mathbf{r} \times \mathbf{n}) \times \mathbf{r}$$

$$= Hr\mathbf{n} - \frac{H}{r}\mathbf{n} \cdot \mathbf{r}\mathbf{r}.$$

If then we put $p = \mathbf{n} \cdot \mathbf{r}$, and substitute the value just found in the formula for $\bar{\mathbf{n}}$, we find

$$\bar{\mathbf{n}} = \frac{2p}{r^2}\mathbf{r} - \mathbf{n} \quad\dots\dots\dots\dots\dots\dots\dots(30).$$

It is clear that p is the perpendicular distance from the centre of inversion to the tangent plane to S, measured in the sense of the unit normal $\mathbf{n}$.

To find the second order magnitudes we need the relations obtained by differentiating (28), namely

$$\left. \begin{aligned} E + \mathbf{r} \cdot \mathbf{r}_{11} &= rr_{11} + r_1{}^2 \\ F + \mathbf{r} \cdot \mathbf{r}_{12} &= rr_{12} + r_1 r_2 \\ G + \mathbf{r} \cdot \mathbf{r}_{22} &= rr_{22} + r_2{}^2 \end{aligned} \right\} \dots\dots\dots\dots\dots(31).$$

The second derivatives $\bar{\mathbf{r}}_{11}$, $\bar{\mathbf{r}}_{12}$, $\bar{\mathbf{r}}_{22}$ are obtained by differentiating (27). On substituting the values so found and making use of (31), we find

$$\left. \begin{aligned} \bar{L} = \bar{\mathbf{n}} \cdot \bar{\mathbf{r}}_{11} &= -\frac{c^2}{r^2}L - \frac{2c^2 p}{r^4}E \\ \bar{M} = \bar{\mathbf{n}} \cdot \bar{\mathbf{r}}_{12} &= -\frac{c^2}{r^2}M - \frac{2c^2 p}{r^4}F \\ \bar{N} = \bar{\mathbf{n}} \cdot \bar{\mathbf{r}}_{22} &= -\frac{c^2}{r^2}N - \frac{2c^2 p}{r^4}G \end{aligned} \right\} \dots\dots\dots\dots(32).$$

From (29) and (32) it follows that

$$\bar{E}\bar{M} - \bar{F}\bar{L} = -\left(\frac{c}{r}\right)^6 (EM - FL),$$

with two similar relations. Hence the differential equation of the lines of curvature on $\bar{S}$ is the same as on S, showing that *lines of curvature invert into lines of curvature*. This is one of the most important properties of inverse surfaces.

83. Curvature. Let the lines of curvature on S be taken as parametric curves, so that $F = M = 0$. It then follows from (29) and (32) that $\bar{F} = \bar{M} = 0$. Hence the parametric curves are lines of curvature on the inverse surface also, affording another proof of the property that lines of curvature invert into lines of curvature. The principal curvatures on S are then given by

$$\kappa_a = \frac{L}{E}, \quad \kappa_b = \frac{N}{G},$$

and those on the inverse surface by

$$\left.\begin{array}{l} \bar{\kappa}_a = \dfrac{\bar{L}}{\bar{E}} = -\dfrac{r^2}{c^2}\kappa_a - \dfrac{2p}{c^2} \\[2mm] \bar{\kappa}_b = \dfrac{\bar{N}}{\bar{G}} = -\dfrac{r^2}{c^2}\kappa_b - \dfrac{2p}{c^2} \end{array}\right\} \quad \dots\dots\dots\dots\dots(33).$$

Hence

$$\bar{\kappa}_a - \bar{\kappa}_b = -\frac{r^2}{c^2}(\kappa_a - \kappa_b),$$

so that *umbilici invert into umbilici*. The specific curvature of the inverse surface is

$$\bar{K} = \bar{\kappa}_a\bar{\kappa}_b = \frac{r^4}{c^4}K + \frac{2r^2 p}{c^4}J + \frac{4p^2}{c^4},$$

and the first curvature is

$$\bar{J} = \bar{\kappa}_a + \bar{\kappa}_b = -\frac{r^2}{c^2}J - \frac{4p}{c^2}.$$

The normal curvature in any direction follows from (33) by Euler's Theorem. Thus

$$\bar{\kappa}_n = \bar{\kappa}_a \cos^2\psi + \bar{\kappa}_b \sin^2\psi$$

$$= -\frac{r^2}{c^2}\kappa_n - \frac{2p}{c^2} \quad \dots\dots\dots\dots\dots\dots\dots(34),$$

since the angle ψ is unaltered by inversion.

The perpendicular from the centre of inversion to the tangent plane to the inverse surface is

$$\bar{p} = \bar{\mathbf{r}} \cdot \bar{\mathbf{n}} = \frac{c^2}{r^2}\mathbf{r} \cdot \left(\frac{2p}{r^2}\mathbf{r} - \mathbf{n}\right),$$

so that

$$\bar{p} = \frac{c^2}{r^2}p.$$

Ex. Show that the quantity $\left(r\kappa_n + \dfrac{p}{r}\right)$ is merely altered in sign by inversion of the surface.

EXAMPLES XI

1. Show that the centres of curvature for the central quadric are, with the notation of Art. 62,

$$\bar{x} = \sqrt{\frac{(a+u)^3(a+v)}{a(a-b)(a-c)}}, \quad \bar{y} = \sqrt{\frac{(b+u)^3(b+v)}{b(b-c)(b-a)}},$$

$$\bar{z} = \sqrt{\frac{(c+u)^3(c+v)}{c(c-a)(c-b)}},$$

and

$$\bar{x}' = \sqrt{\frac{(a+v)^3(a+u)}{a(a-b)(a-c)}}, \quad \bar{y}' = \sqrt{\frac{(b+v)^3(b+u)}{b(b-c)(b-a)}},$$

$$\bar{z}' = \sqrt{\frac{(c+v)^3(c+u)}{c(c-a)(c-b)}}.$$

Hence prove that the two sheets of the centro-surface are identical. Prove also that

$$\frac{a\bar{x}^2}{(a+u)^3} + \frac{b\bar{y}^2}{(b+u)^3} + \frac{c\bar{z}^2}{(c+u)^3} = 0,$$

and

$$\frac{a\bar{x}^2}{(a+u)^2} + \frac{b\bar{y}^2}{(b+u)^2} + \frac{c\bar{z}^2}{(c+u)^2} = 1.$$

The elimination of u between these two equations gives the equation of the centro-surface. (Cf. Forsyth, pp. 113 –115.)

2. The *middle evolute* of a surface, as defined by Ribaucour, is the locus of the point midway between the two centres of curvature. The current point on the middle evolute is therefore given by

$$\bar{r} = r + \tfrac{1}{2}(a+\beta)\,n,$$

where r is a point on the given surface. Find the fundamental magnitudes and the unit normal for the middle evolute.

3. Give a geometrical proof of the theorem (Art. 81) that there is a family of surfaces normal to the tangents to a family of geodesics on a given surface.

4. Calculate the fundamental magnitudes for the complementary surface determined by formula (25), Art. 81.

5. Verify the values of the second order magnitudes for the inverse surface as given by formula (32).

6. Show that conjugate lines are not generally inverted into conjugate lines, nor asymptotic lines into asymptotic lines.

7. Determine the conjugate systems on a surface such that the corresponding curves on a parallel surface form a conjugate system.

8. Determine the character of a surface such that its asymptotic lines correspond to conjugate lines on a parallel surface.

9. The centro-surface of a helicoid is another helicoid with the same axis and pitch as the given surface.

10. A sphere of radius a rolls on the outside of a closed oval surface of volume V and area S; and the parallel surface, which is its outer envelope, has volume V' and area S'. Show that

$$V' - V = a(S' + S) - \tfrac{16}{3}\pi a^3.$$

11. In the previous exercise, the fundamental magnitudes for the outer surface are given by

$$E' = (1 - 4a^2K)E + 4a(aJ - 1)L,$$
$$F' = (1 - 4a^2K)F + 4a(aJ - 1)M,$$
$$G' = (1 - 4a^2K)G + 4a(aJ - 1)N,$$

and

$$L' = (1 - 2aJ)L + 2aKE,$$
$$M' = (1 - 2aJ)M + 2aKF,$$
$$N' = (1 - 2aJ)N + 2aKG.$$

CHAPTER IX

CONFORMAL AND SPHERICAL REPRESENTATIONS.
MINIMAL SURFACES

CONFORMAL REPRESENTATION

84. Conformal representation. When a one-to-one correspondence exists between the points of two surfaces, either surface
may be said to be *represented* on the other. Thus two concentric
spherical surfaces are represented on each other, the two points on
the same radial line corresponding. The surface of a cylinder is
represented on that portion of a plane into which it can be developed. A conical surface is likewise represented on the portion of
a plane into which it can be unwrapped. The surface of a film is
represented on the portion of the screen on which the image is
thrown, a point of the film corresponding to that point of the
screen on which its image appears. Likewise the surface of the
earth is represented on a map, each point of the map corresponding to one and only one point on the earth's surface.

In general, corresponding portions of the two surfaces represented
are not similar to each other. But in the examples mentioned
above there is similarity of the corresponding small elements.
When this relation holds the representation is said to be *conformal*.
The condition necessary for this is clearly that, in the neighbourhood of two corresponding points, all corresponding elements of
arc should be proportional. If this relation holds it follows by
elementary geometry that all corresponding infinitesimal figures
on the two surfaces are similar. Let parameters u, v be chosen to
map out the surfaces S, $\bar{S}$ so that corresponding points on the two
surfaces have the same parameter values. Let the squares of their
linear elements be

$$ds^2 = E\,du^2 + 2F\,du\,dv + G\,dv^2,$$

and $$d\bar{s}^2 = \bar{E}\,du^2 + 2\bar{F}\,du\,dv + \bar{G}\,dv^2.$$

Then, if $d\bar{s}/ds$ has the same value for all directions at a given point,
we must have

$$\frac{\bar{E}}{E} = \frac{\bar{F}}{F} = \frac{\bar{G}}{G} = \frac{d\bar{s}^2}{ds^2} = \eta^2 \quad\ldots\ldots\ldots\ldots(1),$$

where η is a function of u and v or a constant. Conversely, if these relations hold, all corresponding elements of arc at a given point have the same ratio, and the representation is conformal. Then

$$d\bar{s} = \eta \, ds.$$

The quantity η may be called the *linear magnification*. When it has the value unity for all points of the surface, $d\bar{s} = ds$. The conformal representation is then said to be *isometric*, and the two surfaces are said to be *applicable*. In this case corresponding elements of the two surfaces are congruent. In the examples mentioned above the cylindrical and the conical surfaces are applicable to those portions of the plane into which they can be developed.

We may notice in passing that *null lines on a surface correspond to null lines in the conformal representation*. For since $d\bar{s}^2 = \eta^2 ds^2$, if ds^2 vanishes along a curve on S, $d\bar{s}^2$ will vanish along the corresponding curve on $\bar{S}$. Conversely, *if null lines on S correspond to null lines on $\bar{S}$, the representation is conformal*. Let the null lines be taken as parametric curves. Then

$$E = G = 0 \quad \text{and} \quad \bar{E} = \bar{G} = 0.$$

Therefore
$$\frac{d\bar{s}^2}{ds^2} = \frac{2\bar{F}\,du\,dv}{2F\,du\,dv} = \frac{\bar{F}}{F}.$$

Since then $d\bar{s}/ds$ has the same value for all arcs through a given point, the representation is conformal.

It would be out of place here to attempt a systematic discussion of conformal representation. We shall be content with giving the important cases of the representation of a sphere and a surface of revolution on a plane. We may also mention the following general theorem, whose proof depends upon the theory of functions of a complex variable:

If ϕ, ψ are a pair of isometric parameters on the surface S, and u, v isometric parameters on $\bar{S}$, the most general conformal representation of one surface on the other is given by

$$u + iv = f(\phi + i\psi) \quad \ldots \ldots \ldots \ldots \ldots (2),$$

where f is any analytic function of the argument, the point (x, y) corresponding to the point (ϕ, ψ).

85. Surface of revolution. Consider, as an example, a conformal representation of a surface of revolution upon a plane. If the axis of the surface is taken as z-axis, and u is the distance of a point on the surface from this axis, the coordinates of the point may be expressed

$$x = u \cos \phi, \quad y = u \sin \phi, \quad z = f(u),$$

where ϕ is the longitude. The square of the linear element is

$$ds^2 = du^2 + dz^2 + u^2 d\phi^2$$
$$= (1 + f_1^2) du^2 + u^2 d\phi^2.$$

If then we put

$$d\psi = \frac{1}{u} \sqrt{1 + f_1^2} du,$$

we have

$$ds^2 = u^2 (d\psi^2 + d\phi^2).$$

Thus ϕ, ψ are isometric parameters on the surface of revolution. The curves $\phi = \text{const.}$ are the meridians, and the curves $\psi = \text{const.}$ the parallels.

On the plane, rectangular coordinates x, y are isometric parameters, since $d\bar{s}^2 = dx^2 + dy^2$. Consider the representation defined by

$$x + iy = k(\phi + i\psi),$$

that is

$$x = k\phi, \quad y = k\psi \quad \ldots\ldots\ldots\ldots\ldots\ldots (3),$$

where k is constant. Then the point (x, y) on the plane corresponds to the point (ϕ, ψ) on the surface of revolution. Further

$$d\bar{s}^2 = dx^2 + dy^2 = k^2 (d\phi^2 + d\psi^2)$$
$$= \frac{k^2}{u^2} ds^2,$$

showing that the representation is conformal, with a linear magnification k/u. The lines $x = \text{const.}$ correspond to meridians on the surface of revolution, and the lines $y = \text{const.}$ to the parallels.

Any straight line $ax + by + c = 0$ on the plane cuts the lines $x = \text{const.}$ at a constant angle. Therefore, since the representation is conformal, the corresponding line $k(a\phi + b\psi) + c = 0$ on the surface of revolution cuts the meridians at a constant angle. Such a line is called a *loxodrome curve*, or briefly a *loxodrome*, on the surface of revolution. On substituting the value of ψ we find, for the equation of loxodromes on the surface,

$$a\phi + b \int \frac{1}{u} \sqrt{1 + f_1^2} du = \text{const.} \quad \ldots\ldots\ldots\ldots (4).$$

A triangle in the plane corresponds to a curvilinear triangle bounded by loxodromes on the surface of revolution. And, since corresponding angles in the two figures are equal, it follows that *the sum of the angles of a curvilinear triangle, bounded by loxodromes on a surface of revolution, is equal to two right angles.*

Finally we may show that, *when the linear element of a surface is reducible to the form*

$$ds^2 = U\,(du^2 + dv^2) \quad\ldots\ldots\ldots\ldots\ldots(5),$$

where U is a function of u only, or a constant, the surface is applicable to a surface of revolution. For if we write $r = \sqrt{U}$, and solve this equation for u in terms of r, the equation

$$du = \frac{1}{r}\sqrt{1 + f'^2}\,dr$$

determines a function $f(r)$ such that the surface of revolution

$$x = r\cos v, \quad y = r\sin v, \quad z = f(r)$$

has the same linear element

$$ds^2 = U\,(du^2 + dv^2)$$

as the given surface.

The above representation of a surface of revolution on a plane is only a particular case. The general conformal representation of the surface of revolution on a plane is given by

$$x + iy = f(\phi + i\psi) \quad\ldots\ldots\ldots\ldots\ldots(6),$$

where f is any analytic function of the argument.

86. Surface of a sphere. The theory of maps, whether geographical or astronomical, renders the sphere an important example of a surface of revolution. The surface of the earth, or the celestial sphere, is to be represented conformally on a plane, so that there is similarity of detail though not similarity at large. If ϕ is the longitude and λ the latitude, then, with the centre as origin,

$$z = a\sin\lambda, \quad u = a\cos\lambda,$$

a being the radius of the sphere. Thus the square of the linear element is

$$ds^2 = a^2 d\lambda^2 + a^2\cos^2\lambda\,d\phi^2$$
$$= a^2\cos^2\lambda\,(\sec^2\lambda\,d\lambda^2 + d\phi^2).$$

If then we write $\qquad \psi = \log\tan\left(\dfrac{\lambda}{2} + \dfrac{\pi}{4}\right) \quad\ldots\ldots\ldots\ldots(7),$

so that $\qquad\qquad d\psi = \sec\lambda\,d\lambda,$

we have $\qquad\qquad ds^2 = a^2\cos^2\lambda\,(d\phi^2 + d\psi^2) \quad\ldots\ldots\ldots\ldots(8).$

The particular conformal representation given in the previous Art. becomes for the sphere

$$x = k\phi, \quad y = k \log \tan \left(\frac{\lambda}{2} + \frac{\pi}{4}\right).$$

Meridians on the sphere, with a constant difference of longitude, are represented by equidistant parallel straight lines $x = $ const. Parallels of latitude, with a constant difference of latitude, are represented by parallel straight lines $y = $ const., whose distance apart increases toward the poles. The magnification is

$$\eta = \frac{k}{u} = \frac{k}{a \cos \lambda},$$

which increases from k/a at the equator to infinity at the poles. This representation of a sphere on a plane is known as *Mercator's projection*.

Another conformal representation of a sphere on a plane is given by

$$x + iy = ke^{ic(\phi + i\psi)},$$

where c is a constant. This is equivalent to

$$x = ke^{-c\psi} \cos c\phi, \quad y = ke^{-c\psi} \sin c\phi \quad \ldots\ldots\ldots\ldots(9).$$

That the representation is conformal is easily verified. For

$$d\bar{s}^2 = dx^2 + dy^2 = c^2 k^2 e^{-2c\psi} (d\phi^2 + d\psi^2),$$

and therefore, in virtue of (8),

$$d\bar{s}^2 = \frac{c^2 k^2}{a^2} \frac{e^{-2c\psi}}{\cos^2 \lambda} ds^2,$$

as required. The linear magnification is now

$$\eta = \frac{ck}{a} \frac{e^{-c\psi}}{\cos \lambda} = \frac{ck}{a} \frac{(1 - \sin \lambda)^{\frac{1}{2}(c-1)}}{(1 + \sin \lambda)^{\frac{1}{2}(c+1)}} \quad \ldots\ldots\ldots\ldots(10).$$

Meridians on the sphere are represented by the straight lines

$$y = x \tan c\phi,$$

through the origin. Parallels of latitude are represented by the concentric circles

$$x^2 + y^2 = k^2 e^{-2c\psi} = k^2 \left(\frac{1 - \sin \lambda}{1 + \sin \lambda}\right)^c,$$

with centre at the origin. The particular case for which $c = 1$ is known as *stereographic projection*. It is sometimes used for terrestrial maps. Various other values of c are used for star-maps.

SPHERICAL REPRESENTATION

87. Spherical image. We shall now consider briefly the spherical representation of a surface, in which each point or configuration on the surface has its representation on a unit sphere, whose centre may be taken at the origin. If $\mathbf{n}$ is the unit normal at the point P on the surface, the point Q whose position vector is $\mathbf{n}$ is said to correspond to P, or to be the *image* of P. Clearly Q lies on the unit sphere; and if P moves in any curve on the surface, Q moves in the corresponding curve on the sphere.

Since the position vector $\bar{\mathbf{r}}$ of Q is given by

$$\bar{\mathbf{r}} = \mathbf{n},$$

it follows from Art. 27 that

$$\bar{\mathbf{r}}_1 = \mathbf{n}_1 = H^{-2}\{(FM - GL)\,\mathbf{r}_1 + (FL - EM)\,\mathbf{r}_2\},$$
$$\bar{\mathbf{r}}_2 = \mathbf{n}_2 = H^{-2}\{(FN - GM)\,\mathbf{r}_1 + (FM - EN)\,\mathbf{r}_2\}.$$

Consequently, if e, f, g denote the fundamental magnitudes of the first order for the spherical image,

$$e = H^{-2}(EM^2 - 2FLM + GL^2),$$
$$f = H^{-2}(EMN - FM^2 - FLN + GLM),$$
$$g = H^{-2}(EN^2 - 2FMN + GM^2),$$

or, in terms of the first and second curvatures,

$$\left. \begin{aligned} e &= JL - KE \\ f &= JM - KF \\ g &= JN - KG \end{aligned} \right\} \quad \dots\dots\dots\dots\dots(11).$$

Hence also $\qquad eg - f^2 = K^2 H^2,$

which we may write $\qquad h^2 = K^2 H^2$

or $\qquad h = \epsilon K H \quad \dots\dots\dots\dots\dots\dots(12),$

where $\epsilon = \pm 1$ according as the surface is synclastic or anticlastic. The areas of corresponding elements of the spherical image and the given surface are $h\,du\,dv$ and $H\,du\,dv$, and their ratio is therefore numerically equal to K. This property is sometimes used to define the "specific curvature." We may also observe in passing that, since h^2 must be positive and not zero, K must not vanish, so that the surface to be represented on the sphere cannot be a developable surface.

In virtue of (11) we may write the square of the linear element of the image

$$d\bar{s}^2 = J \left(L du^2 + 2M du\, dv + N dv^2\right) - K \left(E du^2 + 2F du\, dv + G dv^2\right),$$

or, if κ is the normal curvature of the given surface in the direction of this arc-element,

$$d\bar{s}^2 = (\kappa J - K)\, ds^2 \quad \dots\dots\dots\dots\dots\dots(13).$$

If then κ_a and κ_b are the principal curvatures of the surface, we may write this, in virtue of Euler's theorem,

$$d\bar{s}^2 = \{(\kappa_a + \kappa_b)(\kappa_a \cos^2 \psi + \kappa_b \sin^2 \psi) - \kappa_a \kappa_b\}\, ds^2$$
$$= (\kappa_a^2 \cos^2 \psi + \kappa_b^2 \sin^2 \psi)\, ds^2 \quad \dots\dots\dots\dots\dots(14).$$

It is clear from either of these formulae that the value of the quotient $d\bar{s}/ds$ depends upon the direction of the arc-element. Hence *in general the spherical image is not a conformal representation.* It is conformal, however, if $\kappa_a = \pm\, \kappa_b$. When $\kappa_a = -\,\kappa_b$ at all points, the first curvature vanishes identically, and the surface is a minimal surface. Thus *the spherical representation of a minimal surface is conformal.*

Moreover it follows from (14) that the turning values of $d\bar{s}/ds$ are given by $\cos\psi = 0$ and $\sin\psi = 0$. Thus *the greatest and least values of the magnification at a point are numerically equal to the principal curvatures.*

88. Other properties. It is easy to show that *the lines of curvature on a surface are orthogonal in their spherical representation.* For if they are taken as parametric curves we have $F = M = 0$; hence $f = 0$ which proves the statement. Further if $F = 0$ and $f = 0$ we must also have $M = 0$ unless J vanishes identically. Thus, *if the surface is not a minimal surface, the lines of curvature are the only orthogonal system whose spherical image is orthogonal.*

Moreover, *the tangent to a line of curvature is parallel to the tangent to its spherical image at the corresponding point; and, conversely, if this relation holds for a curve on the surface it must be a line of curvature.* For, by Rodrigues' formula (Art. 30), along a line of curvature $d\mathbf{r}$ is parallel to $d\mathbf{n}$ and therefore also to $d\bar{\mathbf{r}}$. Hence the first part of the theorem. Conversely if $d\mathbf{r}$ is parallel to $d\bar{\mathbf{r}}$ it is also parallel to $d\mathbf{n}$. The three vectors $\mathbf{n}$, $\mathbf{n} + d\mathbf{n}$, $d\mathbf{r}$ are therefore coplanar, and the line is a line of curvature.

Again, if dr and ds are two infinitesimal displacements on a given surface, and dn the change in the unit normal due to the former, the directions of the displacements will be conjugate provided

$$dn \cdot ds = 0.$$

And conversely this relation holds if the directions are conjugate. But $dn = d\bar{r}$, where $d\bar{r}$ is the spherical image of dr. Consequently

$$d\bar{r} \cdot ds = 0.$$

Thus, *if two directions are conjugate at a point on a given surface, each is perpendicular to the spherical image of the other at the corresponding point.* It follows that the inclination of two conjugate directions is equal, or supplementary, to that of their spherical representations.

Further, an asymptotic line is self-conjugate. Hence *an asymptotic line on a surface is perpendicular to its spherical image at the corresponding point.*

Ex. 1. Taking the lines of curvature as parametric curves, deduce the theorem that a line of curvature is parallel to its spherical image at the corresponding point, from the formulae for $\bar{r}_1$ and $\bar{r}_2$ in Art. 87.

Ex. 2. Prove otherwise that the inclination of conjugate lines is equal or supplementary to that of their spherical image.

Let the conjugate lines be taken as parametric curves, so that $M = 0$. Then equations (11) become

$$e = \frac{GL^2}{H^2}, \quad f = -\frac{FLN}{H^2}, \quad g = \frac{EN^2}{H^2}.$$

Hence the angle Ω between the parametric curves on the unit sphere is given by

$$\cos \Omega = \frac{f}{\sqrt{eg}} = \mp \frac{F}{\sqrt{EG}} = \mp \cos \omega,$$

the negative or the positive sign being taken according as the surface is synclastic or anticlastic.

Ex. 3. If a line of curvature is plane, its plane cuts the surface at a constant angle.

If r is a point on the line of curvature, n is the corresponding point on the spherical image. But we have seen that the tangents to these at corresponding points are parallel; and therefore

$$\frac{dn}{ds} = \pm \frac{dr}{ds}.$$

Let **a** be the (constant) unit normal to the plane of the curve. Then

$$\frac{d}{ds}(\mathbf{a}\cdot\mathbf{n})=\mathbf{a}\cdot\frac{d\mathbf{n}}{d\bar{s}}\frac{d\bar{s}}{ds}=\pm\,\mathbf{a}\cdot\frac{d\mathbf{r}}{ds}\frac{d\bar{s}}{ds}.$$

But $\dfrac{d\mathbf{r}}{ds}$ lies in the given plane and is therefore perpendicular to **a**. Thus the last expression vanishes, showing that $\mathbf{a}\cdot\mathbf{n}$ is constant, so that the plane cuts the surface at a constant angle.

Moreover, the relation $\mathbf{a}\cdot\mathbf{n}=\text{const.}$ is equivalent to $\mathbf{a}\cdot\bar{\mathbf{r}}=\text{const.}$ Thus the projection of $\bar{\mathbf{r}}$ on the diameter parallel to **a** is constant, showing that the spherical image of a plane line of curvature is a small circle, whose plane is parallel to that of the line of curvature.

89. Second order magnitudes. We may also calculate the magnitudes of the second order, $\bar{L}$, $\bar{M}$, $\bar{N}$, for the spherical representation. Its unit normal $\bar{\mathbf{n}}$ is given by

$$h\bar{\mathbf{n}}=\bar{\mathbf{r}}_1\times\bar{\mathbf{r}}_2=\mathbf{n}_1\times\mathbf{n}_2=KH\mathbf{n},$$

in virtue of Art. 27 (18). But $h=\epsilon KH$, and therefore

$$\bar{\mathbf{n}}=\epsilon\mathbf{n}\ \dotfill(15),$$

where $\epsilon=\pm1$ according as the surface is synclastic or anticlastic. Consequently

$$\bar{L}=\bar{\mathbf{n}}\cdot\bar{\mathbf{r}}_{11}=\epsilon\mathbf{n}\cdot\mathbf{n}_{11}=\epsilon\left\{\frac{\partial}{\partial u}(\mathbf{n}\cdot\mathbf{n}_1)-\mathbf{n}_1{}^2\right\}$$

$$=-\epsilon\mathbf{n}_1{}^2=-\epsilon e.$$

Proceeding in like manner for the others we have

$$\left.\begin{aligned}\bar{L}&=-\epsilon e\\ \bar{M}&=-\epsilon f\\ \bar{N}&=-\epsilon g\end{aligned}\right\}\ \dotfill(16).$$

Thus only the first order magnitudes need be considered. Also the radius of curvature of any normal section of the sphere is given by

$$\rho=\frac{e\,du^2+2f\,du\,dv+g\,dv^2}{\bar{L}\,du^2+2\bar{M}\,du\,dv+\bar{G}\,dv^2},$$

and is therefore numerically equal to unity, as we should expect.

*90. Tangential coordinates. The tangential coordinates of a point P on a given surface are the direction cosines of the normal at P, and the perpendicular distance of the origin from the tangent plane at that point. These are equivalent to the unit normal **n** at P, and the distance p from the origin to the tangent plane, measured in the direction of $\mathbf{n}$. If $\mathbf{r}$ is the position vector of P, we have

$$p = \mathbf{r} \cdot \mathbf{n}$$

and, on differentiation with respect to u and v, it follows that

$$\left.\begin{array}{l} p_1 = \mathbf{r} \cdot \mathbf{n}_1 \\ p_2 = \mathbf{r} \cdot \mathbf{n}_2 \end{array}\right\}.$$

The position vector $\mathbf{r}$ of the current point on the surface may be expressed in terms of $\mathbf{n}, \mathbf{n}_1, \mathbf{n}_2$. For, by the usual formula for the expression of a vector in terms of three non-coplanar vectors,

$$[\mathbf{n}, \mathbf{n}_1, \mathbf{n}_2]\,\mathbf{r} = [\mathbf{r}, \mathbf{n}_1, \mathbf{n}_2]\,\mathbf{n} + [\mathbf{r}, \mathbf{n}_2, \mathbf{n}]\,\mathbf{n}_1 + [\mathbf{r}, \mathbf{n}, \mathbf{n}_1]\,\mathbf{n}_2.$$

Now in Art. 27 it was shown that

$$[\mathbf{n}, \mathbf{n}_1, \mathbf{n}_2] = HK.$$

Further $\mathbf{n}_1 \times \mathbf{n}_2 = HK\mathbf{n}$,

so that $[\mathbf{r}, \mathbf{n}_1, \mathbf{n}_2] = HKp$.

Again $\mathbf{n}_2 \times \mathbf{n} = \dfrac{1}{HK}(g\mathbf{n}_1 - f\mathbf{n}_2)$,

and therefore $[\mathbf{r}, \mathbf{n}_2, \mathbf{n}] = \dfrac{1}{HK}(gp_1 - fp_2)$.

Similarly $[\mathbf{r}, \mathbf{n}, \mathbf{n}_1] = \dfrac{1}{HK}(ep_2 - fp_1)$.

On substitution of these values in the above formula we have

$$\mathbf{r} = p\mathbf{n} + \frac{1}{h^2}\{(gp_1 - fp_2)\,\mathbf{n}_1 + (ep_2 - fp_1)\,\mathbf{n}_2\} \quad \ldots\ldots(17).$$

Hence, *when p and $\mathbf{n}$ are given and their derivatives can be calculated, the surface is completely determined.*

MINIMAL SURFACES

91. General properties. A *minimal surface* may be defined as one whose first curvature, J, vanishes identically. Thus the principal curvatures at any point of the surface are equal in magnitude and opposite in sign, and the indicatrix is a rectangular hyperbola. Hence the asymptotic lines form an orthogonal system, bisecting the angles between the lines of curvature. The vanishing of the first curvature is expressed by the equation

$$EN - 2FM + GL = 0 \quad \ldots\ldots\ldots\ldots\ldots(18),$$

which is satisfied by all minimal surfaces.

Minimal surfaces derive their name from the fact that they are surfaces of *minimum area* satisfying given boundary conditions. They are illustrated by the shapes of thin soap films in equilibrium, with the air pressure the same on both sides. If this property of least area be taken as defining minimal surfaces, the use of the Calculus of Variations leads to the vanishing of the first curvature as an equivalent property.

We have seen that *the asymptotic lines on a minimal surface form an orthogonal system*. This follows also from (18). For, if these lines are taken as parametric curves we have $L = N = 0$, while M does not vanish. Hence $F = 0$, showing that the parametric curves are orthogonal. Conversely, *if the asymptotic lines are orthogonal, the surface is minimal*. For, with the same choice of parametric curves we have $L = N = 0$ and $F = 0$, so that J vanishes identically.

Further, *the null lines on a minimal surface form a conjugate system*. For, if these are taken as parametric curves we have $E = G = 0$, while F does not vanish. It therefore follows from (18) that $M = 0$, showing that the parametric curves are conjugate. Conversely, *if the null lines are conjugate the surface is minimal*. For then, with the same choice of parametric curves, $E = G = 0$ and $M = 0$, so that (18) is satisfied identically.

Again, *the lines of curvature on a minimal surface form an isometric system*. To prove this let the lines of curvature be taken as parametric curves. Then $F = 0$ and $M = 0$, and the equation (18) becomes

$$\frac{L}{E} + \frac{N}{G} = 0,$$

while the Mainardi-Codazzi relations reduce to

$$L_2 = \frac{1}{2}\left(\frac{L}{E} + \frac{N}{G}\right) E_2 = 0,$$

and

$$N_1 = \frac{1}{2}\left(\frac{L}{E} + \frac{N}{G}\right) G_1 = 0.$$

Thus L is a function of u only, and N a function of v only. Consequently

$$\frac{\partial^2}{\partial u \partial v} \log \frac{E}{G} = \frac{\partial^2}{\partial u \partial v} \log \left(-\frac{L}{N}\right) = 0,$$

showing that the parametric curves are isometric.

92. Spherical image. The fundamental magnitudes for the spherical representation, as given by (11), become in the case of a minimal surface

$$e = -KE, \quad f = -KF, \quad g = -KG \quad \ldots\ldots\ldots(19).$$

From these relations several interesting general properties may be deduced.

We have already seen that *the spherical representation of a minimal surface is conformal.* This also follows directly from (19). For, in virtue of these relations,

$$d\bar{s}^2 = -Kds^2.$$

Thus $d\bar{s}/ds$ is independent of the direction of the arc-element through the point, and the representation is conformal. The magnification has the value $\sqrt{-K}$, the second curvature being essentially negative for a real minimal surface. The converse of the above theorem has already been considered in Art. 87, where it was shown that if the spherical image of a surface is a conformal representation, either the surface is minimal, or else its principal curvatures are equal at each point.

Further, *null lines on a minimal surface become both null lines and asymptotic lines in the spherical representation.* For, if the null lines be taken as parametric curves, we have

$$E = 0, \quad G = 0,$$

and therefore, by (19), $\quad e = 0, \quad g = 0.$

Thus the parametric curves in the spherical image also are null lines. Again, considering the second order magnitudes for the sphere, we have

$$\overline{L} = -\epsilon e = 0,$$
$$\overline{N} = -\epsilon g = 0,$$

and therefore the parametric curves in the spherical image are also asymptotic lines.

Conversely, *if the null lines on a surface become null lines in the spherical representation, either the surface is minimal, or else its principal curvatures are equal.* To prove this theorem, take the null lines as parametric curves. Then $E = G = 0$; and since the parametric curves are also null lines in the spherical image, $e = g = 0.$ But for any surface

$$e = JL - KE \atop g = JN - KG \Big\},$$

so that we must have

$$JL = 0 \quad \text{and} \quad JN = 0.$$

Consequently either $J = 0$ and the surface is minimal, or else

$$L = 0 \quad \text{and} \quad N = 0.$$

In the latter case it follows that

$$J = \frac{2M}{F} \quad \text{and} \quad K = \frac{M^2}{F^2},$$

and therefore $\qquad\qquad J^2 - 4K = 0$

which is the condition that the principal curvatures should be equal.

Lastly, *isometric lines on a minimal surface are also isometric in the spherical representation.* This is obvious from the fact that the spherical image is a conformal representation. It may also be proved as follows. If the isometric lines are taken as parametric curves we have

$$F = 0, \quad \frac{E}{G} = \frac{U}{V},$$

where U is a function of u only and V a function of v only. From (19) it then follows that

$$f = 0, \quad \frac{e}{g} = \frac{E}{G} = \frac{U}{V},$$

showing that the parametric curves in the spherical image are also isometric. Hence the theorem. In particular the spherical image of the lines of curvature on a minimal surface are isometric curves, for the lines of curvature on a minimal surface have been shown to be isometric.

93. Cartesian coordinates. If we use the form $z = f(x, y)$ for the equation of a surface, the differential equation (18) of minimal surfaces becomes

$$\left\{1 + \left(\frac{\partial z}{\partial y}\right)^2\right\}\frac{\partial^2 z}{\partial x^2} - 2\frac{\partial z}{\partial x}\frac{\partial z}{\partial y}\frac{\partial^2 z}{\partial x \partial y} + \left\{1 + \left(\frac{\partial z}{\partial x}\right)^2\right\}\frac{\partial^2 z}{\partial y^2} = 0 \quad (20).$$

This form of the equation is useful for particular problems.

By way of illustration we may prove that *the catenoid is the only minimal surface of revolution.* If the axis of revolution is taken as the z-axis we may write

$$z = f(x^2 + y^2),$$

12—2

where the form of the function f is to be determined so that the surface may be minimal. By differentiation we have

$$\frac{\partial z}{\partial x} = 2xf', \quad \frac{\partial z}{\partial y} = 2yf',$$

$$\frac{\partial^2 z}{\partial x^2} = 2f' + 4x^2 f'', \quad \frac{\partial^2 z}{\partial x \partial y} = 4xy f'', \text{ etc.},$$

and on substituting these values in (20) we have

$$(x^2 + y^2)f'' + 2(x^2 + y^2)f'^3 + f' = 0.$$

On putting $r^2 = x^2 + y^2$ we may write this equation

$$r\frac{d^2 z}{dr^2} + \left\{1 + \left(\frac{dz}{dr}\right)^2\right\}\frac{dz}{dr} = 0,$$

which gives, on integration,

$$\frac{r\dfrac{dz}{dr}}{\sqrt{1 + \left(\dfrac{dz}{dr}\right)^2}} = a,$$

where a is a constant. A second integration leads to

$$z + c = a \cosh^{-1}\frac{r}{a},$$

or

$$r = a \cosh\frac{z + c}{a}.$$

Thus the only minimal surface of revolution is that formed by the revolution of a catenary about its directrix.

Ex. 1. The only minimal surface of the type

$$z = f(x) + F(y)$$

is the surface

$$az = \log \cos ax - \log \cos ay.$$

On substituting the above expression for z in (20) we find

$$\frac{f''}{1 + f'^2} + \frac{F''}{1 + F'^2} = 0.$$

The first part is a function of x only, and the second a function of y only. Hence each must be constant; so that

$$\frac{f''}{1 + f'^2} = a \text{ and } \frac{F''}{1 + F'^2} = -a.$$

Integration leads to

$$f(x) = \frac{1}{a}\log \cos ax; \quad F(y) = -\frac{1}{a}\log \cos ay.$$

Hence the theorem.

Ex. 2. Show that the surface

$$\sin az = \sinh ax \sinh ay$$

is minimal.

EXAMPLES XII

1. Show that the surfaces
$$x = u \cos \phi, \quad y = u \sin \phi, \quad z = c\phi,$$
and
$$x = u \cos \phi, \quad y = u \sin \phi, \quad z = c \cosh^{-1} \frac{u}{c}$$
are applicable.

2. Show that, in a star-map (Art. 86), the magnification is least on the parallel of latitude $\sin^{-1} c$.

3. Show that rhumb lines of the meridians of a sphere become straight lines in Mercator's projection, and equiangular spirals in a stereographic projection.

4. Find the loxodrome curves on the surfaces in Ex. 1.

5. Find the surface of revolution for which
$$ds^2 = du^2 + (a^2 - u^2)\, d\phi^2.$$

6. Show that, for the surface generated by the revolution of the evolute of the catenary about the directrix, the linear element is reducible to the form
$$ds^2 = du^2 + u\, dv^2.$$

7. Any two stereographic projections of a sphere are inverses of each other, the origin of inversion in either being the origin of projection in the other.

8. In any representation of a surface S on another, S', the cross-ratio of four tangents at a point of S is equal to the cross-ratio of the corresponding tangents to S'.

9. Determine $f(v)$ so that the conoid
$$x = u \cos v, \quad y = u \sin v, \quad z = f(v)$$
may be applicable to a surface of revolution.

10. If the curve of intersection of a sphere and a surface be a line of curvature on the latter, the sphere cuts the surface at a constant angle.

11. If e, f, g refer to the spherical image, prove the formulae
$$h^2 E = e M^2 - 2f LM + g L^2,$$
$$h^2 F = e MN - f(LN + M^2) + g LM,$$
$$h^2 G = e N^2 - 2f MN + g M^2.$$

12. What are the first and second curvatures for the spherical image?

13. The angles between the asymptotic directions at a point on a surface and between their spherical representations are equal or supplementary, according as the second curvature at the point is positive or negative.

14. The osculating planes of a line of curvature and of its spherical image at corresponding points are parallel.

15. Show that the lines of curvature on a surface are given by
$$(eM - fL)\, du^2 + (eN - gL)\, du\, dv + (fN - gM)\, dv^2 = 0,$$
and the principal curvatures by
$$T^2 \kappa^2 - (eN - 2fM + gL)\, \kappa + h^2 = 0.$$

16. The angle θ between any direction on a surface and its spherical image is given by

$$\cos \theta = -\frac{L\,du^2 + 2M\,du\,dv + N\,dv^2}{ds\,d\bar{s}}.$$

Hence an asymptotic direction is perpendicular to its spherical image.

17. The formulae (17) of Art. 27 may be written

$$h^2\mathbf{r}_1 = (fM - gL)\,\mathbf{n}_1 + (fL - eM)\,\mathbf{n}_2,$$
$$h^2\mathbf{r}_2 = (fN - gM)\,\mathbf{n}_1 + (fM - eN)\,\mathbf{n}_2.$$

18. Show that the lines of curvature of a surface of revolution remain isometric in their spherical representation.

19. Show that the spherical images of the asymptotic lines on a minimal surface, as well as the asymptotic lines themselves, are an isometric system.

20. If one system of asymptotic lines on a surface are represented on the sphere by great circles, the surface is ruled.

21. The right helicoid is the only real ruled minimal surface.

22. The parameters of the lines of curvature of a minimal surface may be so chosen that the linear elements of the surface and of its spherical image have the respective forms

$$ds^2 = \frac{1}{\kappa}\,(du^2 + dv^2), \quad d\bar{s}^2 = \kappa\,(du^2 + dv^2),$$

where κ is the absolute value of each principal curvature.

23. Prove that Ex. 22 is still true if we write "asymptotic lines" in place of "lines of curvature."

24. Every helicoid is applicable to some surface of revolution, and helices on the former correspond to parallels on the latter.

25. If the fundamental magnitudes of the first order are functions of a single parameter, the surface is applicable to a surface of revolution.

26. Show that the helicoid

$$x = u\cos v, \quad y = u\sin v, \quad z = cv + c\int \sqrt{\frac{u^2 + c^2}{u^2 - c^2}}\,\frac{du}{u}$$

is a minimal surface.

27. Prove that each sheet of the evolute of a pseudo-sphere is applicable to a catenoid.

28. Prove that the surface

$$x = u\cos a + \sin u \cosh v,$$
$$y = v + \cos a \cos u \sinh v,$$
$$z = \sin a \cos u \cosh v$$

is a minimal surface, that the parametric curves are plane lines of curvature, and that the second curvature is

$$-\sin^2 a/(\cosh v + \cos a \cos u)^4.$$

CHAPTER X

CONGRUENCES OF LINES

RECTILINEAR CONGRUENCES

94. Congruence of straight lines. A rectilinear congruence is a two-parameter system of straight lines, that is to say, a family of straight lines whose equation involves two independent parameters. The congruence therefore comprises a double infinitude of straight lines. Such a system is constituted by the normals to a given surface. In dealing with this particular congruence we may take the two parameters as the current parameters u, v for the surface. The normals along any one parametric curve $u = a$ constitute a single infinitude of straight lines, and the whole system of normals a double infinitude. These normals are also normals to the family of surfaces parallel to the given surface, and are therefore termed a *normal* congruence. In general, however, the lines of a rectilinear congruence do not possess this property of normality to a family of surfaces. As other examples of congruences may be mentioned the family of straight lines which intersect two given curves, and the family which intersect a given curve and are tangents to a given surface.

A rectilinear congruence may be represented analytically by an equation of the form

$$\mathbf{R} = \mathbf{r} + t\mathbf{d} \quad\quad\quad\quad\quad\quad\quad\quad (1),$$

where $\mathbf{r}$ and $\mathbf{d}$ are functions of two independent parameters u, v. The point $\mathbf{r}$ may be taken as a point on a surface of reference, or *director surface*, S, which is cut by all the lines of the congruence. We may take $\mathbf{d}$ as a *unit* vector giving the direction of the line or *ray*, and t is then the distance from the director surface to the current point $\mathbf{R}$ on the ray.

We may make a *spherical representation* of the congruence by drawing radii of a unit sphere parallel to the rays of the congruence. Thus the point $\mathbf{d}$ on the sphere represents the ray (1). The linear element $d\sigma$ of the spherical representation is given by

$$d\sigma^2 = (d\mathbf{d})^2 = e\,du^2 + 2f\,du\,dv + g\,dv^2 \quad\quad\quad\quad (2),$$

where $\quad\quad\quad e = \mathbf{d}_1^2, \quad f = \mathbf{d}_1 \cdot \mathbf{d}_2, \quad g = \mathbf{d}_2^2,$

these being the fundamental magnitudes of the first order for the spherical representation. And, since $\mathbf{d}$ is the unit normal to the sphere, we have

$$\mathbf{d} = \frac{\mathbf{d}_1 \times \mathbf{d}_2}{h} \quad\dots\dots\dots\dots\dots\dots\dots(3),$$

where, as usual, $h^2 = eg - f^2.$

Another quadratic form, whose coefficients play an important part in the following argument, is that which arises from the expansion of $d\mathbf{r} \cdot d\mathbf{d}$. We may write this

$$d\mathbf{r} \cdot d\mathbf{d} = (\mathbf{r}_1 du + \mathbf{r}_2 dv) \cdot (\mathbf{d}_1 du + \mathbf{d}_2 dv)$$
$$= a\,du^2 + (b + b')\,du\,dv + c\,dv^2 \quad\dots\dots\dots\dots(4),$$

where $a = \mathbf{r}_1 \cdot \mathbf{d}_1, \quad b = \mathbf{r}_2 \cdot \mathbf{d}_1, \quad b' = \mathbf{r}_1 \cdot \mathbf{d}_2, \quad c = \mathbf{r}_2 \cdot \mathbf{d}_2.$

The rays through any curve C on the director surface form a ruled surface Σ. A relation between the parameters u, v determines such a curve and therefore also a ruled surface. The infinitude of such surfaces, corresponding to the infinitude of relations that may connect u and v, are called the *surfaces of the congruence*. We say that each of these surfaces "passes through" each of the rays that lie upon it. Any surface of the congruence is represented by a curve on the unit sphere, which may be called its spherical representation in the above sense. This curve is the locus of the points on the sphere which represent the rays lying upon that surface.

95. Limits. Principal planes. Consider a curve C on the director surface, and the corresponding ruled surface Σ. Let $\mathbf{r}$ and

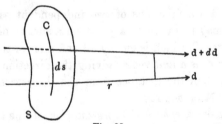

Fig. 23.

$\mathbf{r} + d\mathbf{r}$ be consecutive points on the curve, through which pass the consecutive rays with directions $\mathbf{d}$ and $\mathbf{d} + d\mathbf{d}$ determined by the parameter values u, v and $u + du, v + dv$ respectively. Further let s be the arc-length of the curve C up to the point $\mathbf{r}$, ds the element of arc between the consecutive rays, and let dashes denote derivatives with respect to s. Then the distance r from the director surface to the foot of the common perpendicular to the consecutive rays, as found in Art. 69, is given by

$$r = -\frac{\mathbf{r}' \cdot \mathbf{d}'}{\mathbf{d}'^2} = -\frac{d\mathbf{r} \cdot d\mathbf{d}}{d\mathbf{d}^2}$$

$$= -\left(\frac{a\,du^2 + (b+b')\,du\,dv + c\,dv^2}{e\,du^2 + 2f\,du\,dv + g\,dv^2}\right) \dots\dots(5).$$

The point of the ray determined by r is the central point of the ray relative to the surface Σ.

The distance r from the director surface to the central point is a function of the ratio $du : dv$, so that it varies with the direction of the curve C through the point $\mathbf{r}$. There are two values of this ratio for which r is a maximum or minimum. These are obtained by equating to zero the derivatives of r with respect to du/dv. This leads to the equation

$$[2fa - e(b+b')]\,du^2 + 2(ga - ec)\,du\,dv$$
$$+ [g(b+b') - 2fc]\,dv^2 = 0 \dots\dots(6),$$

which gives the two directions for stationary values of r. To determine these values we have only to eliminate du/dv from the last two equations, thus obtaining the quadratic

$$h^2 r^2 + [ec - f(b+b') + ga]\,r + ac - \tfrac{1}{4}(b+b')^2 = 0 \dots(7),$$

whose roots are the two stationary values required. Denoting these by r_1 and r_2 we have

$$\left.\begin{array}{l} h^2(r_1 + r_2) = f(b+b') - ec - ga \\ 4h^2 r_1 r_2 = 4ac - (b+b')^2 \end{array}\right\} \dots\dots\dots(8).$$

The points on the ray determined by these values of r are called its *limits*. They are the boundaries of the segment of the ray containing the feet of the common perpendiculars to it and the consecutive rays of the congruence. The two ruled surfaces of the congruence which pass through the given ray, and are determined by (6), are called the *principal surfaces* for that ray. Their tangent planes at the limits contain the given ray and the common perpendiculars to it and the consecutive rays of the surfaces. These tangent planes are called the *principal planes* of the ray. They are the central planes of the ray relative to the principal surfaces (Art. 71).

Ex. 1. Find the limits and the principal planes for a ray in the congruence of straight lines which intersect a given circle and its axis.

Let the centre of the given circle be taken as origin, its plane as the director surface and its axis as the z-axis. Let a ray of the congruence meet

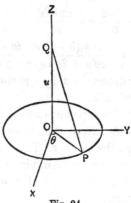

Fig. 24.

the circumference of the circle in P, and the axis in Q. Denote OQ by u and the angle XOP by θ, and let u, θ be taken as parameters for the congruence. Then, if R is the radius of the circle, the position vector of P is

$$\mathbf{r} = (R\cos\theta, \ R\sin\theta, \ 0),$$

and the unit vector $\mathbf{d}$ in the direction PQ of the ray is

$$\mathbf{d} = \frac{(-R\cos\theta, \ -R\sin\theta, \ u)}{\sqrt{R^2 + u^2}}.$$

From these it is easy to verify that

$$e = \frac{R^2}{(R^2 + u^2)^{\frac{3}{2}}}, \quad f = 0, \quad g = \frac{R^2}{R^2 + u^2},$$

while
$$a = 0, \quad b = 0, \quad b' = 0, \quad c = \frac{-R^2}{\sqrt{R^2 + u^2}}.$$

The equation (7) for the distance of the limits from the point P reduces to

$$r^2 - \sqrt{R^2 + u^2}\, r = 0,$$

so that
$$r_1 = 0, \quad r_2 = \sqrt{R^2 + u^2}.$$

The limits are therefore the points P and Q.

The differential equation (6) for the principal surface becomes simply

$$du\, d\theta = 0.$$

Thus the principal plane through the ray corresponding to $d\theta = 0$ is the plane POQ containing the ray and the axis of the circle. The principal plane corresponding to $du = 0$ is the plane containing PQ and a consecutive ray through Q. These principal planes are clearly perpendicular. The principal surfaces are the planes through the axis, and the cones with vertices on the axis and generators passing through the given circle.

Ex. 2. Examine the congruence of tangents to a given sphere from points on a given diameter.

Take the centre of the sphere as origin and its surface as director surface. Let the given diameter be taken as z-axis and polar axis, the colatitude θ being measured from OZ and the longitude ϕ from the plane ZOX. If a tangent from the point Q on the given diameter touch the sphere at $P(\theta, \phi)$, the position vector of P may be expressed as

$$\mathbf{r} = R\,(\sin \theta \cos \phi, \ \sin \theta \sin \phi, \ \cos \theta),$$

R being the radius of the sphere. The unit vector $\mathbf{d}$ in the direction PQ of the ray is

$$\mathbf{d} = (-\cos \theta \cos \phi, \ -\cos \theta \sin \phi, \ \sin \theta).$$

Taking θ, ϕ as parameters, show that

$$e = 1, \quad f = 0, \quad g = \cos^2 \theta,$$

and

$$a = 0, \quad b = b' = 0, \quad c = -R \sin \theta \cos \theta.$$

Hence show that the distances of the limits from P are

$$r_1 = 0, \quad r_2 = R \tan \theta,$$

so that P and Q are the limits. The equation (6) becomes

$$d\theta\, d\phi = 0.$$

Hence the principal surfaces are the planes through the given diameter and the tangent cones from points on that diameter.

96. Hamilton's formula. Let the parameters be so chosen that the principal surfaces correspond to the parametric curves. Then the equation (6) determining the principal surfaces must be equivalent to $du\,dv = 0$. This requires

$$\left.\begin{array}{r} 2fa - e\,(b + b') = 0 \\ 2fc - g\,(b + b') = 0 \end{array}\right\},$$

and therefore, since the coefficients of the two quadratic forms are not proportional, we must have

$$f = 0, \quad b + b' = 0 \quad \dots\dots\dots\dots\dots(9).$$

The first of these is equivalent to $\mathbf{d}_1 \cdot \mathbf{d}_2 = 0$. Hence *the principal surfaces are represented on the unit sphere by orthogonal curves.*

The mutual perpendicular to the consecutive rays $\mathbf{d}$ and $\mathbf{d} + d\mathbf{d}$ is perpendicular to both $\mathbf{d}$ and $d\mathbf{d}$. Hence it is perpendicular to $\mathbf{d}$ and $\dfrac{d\mathbf{d}}{d\sigma}$, where $d\sigma$ is the arc element of the unit sphere corresponding to $d\mathbf{d}$. But these are two unit vectors, perpendicular to each other. Hence the unit vector in the direction of the common perpendicular is $\dfrac{d\mathbf{d}}{d\sigma} \times \mathbf{d}$. Now, in virtue of (3) we may write this

$$\left(\mathbf{d}_1 \frac{du}{d\sigma} + \mathbf{d}_2 \frac{dv}{d\sigma}\right) \times \frac{\mathbf{d}_1 \times \mathbf{d}_2}{h},$$

or, since $\mathbf{d}_1$ and $\mathbf{d}_2$ are perpendicular ($f = 0$),

$$\frac{1}{d\sigma \sqrt{eg}} (g\mathbf{d}_1\, dv - e\mathbf{d}_2\, du) \dots\dots\dots\dots(10).$$

The consecutive rays corresponding to the limits are determined by $dv = 0$ and $du = 0$, that is by the parametric curves. For these we have respectively $d\sigma = \sqrt{e}\, du$ and $d\sigma = \sqrt{g}\, dv$; so that (10) becomes in the cases corresponding to the limits

$$-\frac{1}{\sqrt{g}}\, \mathbf{d}_2 \text{ and } \frac{1}{\sqrt{e}}\, \mathbf{d}_1 \dots\dots\dots\dots(11),$$

which are perpendicular to each other. Now the tangent plane to the ruled surface of the congruence through the consecutive rays is parallel to $\mathbf{d}$ and the common perpendicular. Therefore, in virtue of (11), *the two principal planes for any ray are perpendicular to each other.*

The angle θ, between the common perpendicular (10) in the general case and that corresponding to the principal surface $dv = 0$, is given by

$$\cos\theta = -\frac{1}{\sqrt{g}}\, \mathbf{d}_2 \cdot \frac{(g\mathbf{d}_1\, dv - e\mathbf{d}_2\, du)}{d\sigma \sqrt{eg}} = \frac{\sqrt{e}\, du}{d\sigma}.$$

Thus
$$\cos^2\theta = \frac{e\, du^2}{e\, du^2 + g\, dv^2},$$

and therefore
$$\sin^2\theta = \frac{g\, dv^2}{e\, du^2 + g\, dv^2}.$$

Further, with this choice of parameters, the distance r from the director surface to the foot of the common perpendicular is given by

$$r = -\frac{a\, du^2 + c\, dv^2}{e\, du^2 + g\, dv^2},$$

and therefore for the limits ($dv = 0$ and $du = 0$),

$$r_1 = -\frac{a}{e}, \quad r_2 = -\frac{c}{g}.$$

It follows immediately that

$$r = r_1 \cos^2 \theta + r_2 \sin^2 \theta \quad\ldots\ldots\ldots\ldots\ldots(12).$$

This is *Hamilton's formula* connecting the position of the central point of a ray, relative to any surface through it, with the inclination of the central plane of the ray for this surface to the principal planes of the ray. The formula is independent of the choice of parameters. Also, if the director surface is changed, the three distances r, r_1, r_2 are altered by the same amount, and the formula still holds.

We may observe in passing that the normal to a surface of the congruence through the ray $\mathbf{d}$ is perpendicular both to this ray and to the common perpendicular $\dfrac{d\mathbf{d}}{d\sigma} \times \mathbf{d}$ to it and a consecutive ray. It is therefore parallel to

$$\mathbf{d} \times \left(\frac{d\mathbf{d}}{d\sigma} \times \mathbf{d}\right),$$

which is identical with $\dfrac{d\mathbf{d}}{d\sigma}$. Thus *the normal to a surface of the congruence is parallel to the tangent to the spherical representation of the surface*, in the sense of Art. 94.

Ex. For any choice of parametric curves the unit vector perpendicular to consecutive rays is

$$\frac{d\mathbf{d}}{d\sigma} \times \mathbf{d} = \left(\mathbf{d}_1 \frac{du}{d\sigma} + \mathbf{d}_2 \frac{dv}{d\sigma}\right) \times \frac{\mathbf{d}_1 \times \mathbf{d}_2}{h}$$

$$= \frac{1}{h}\left[(f\mathbf{d}_1 - e\mathbf{d}_2)\frac{du}{d\sigma} + (g\mathbf{d}_1 - f\mathbf{d}_2)\frac{dv}{d\sigma}\right].$$

97. Foci. Focal planes. The ruled surface Σ of the congruence will be a *developable* if consecutive generators intersect. The locus of the point of intersection of consecutive rays on the surface is the edge of regression of the developable. It is touched by each of the generators; and the point of contact is called the *focus* of the ray.

Let ρ be the distance of the focus along the ray from the director surface. Then the focus is the point

$$\mathbf{R} = \mathbf{r} + \rho\mathbf{d}.$$

But since the ray touches the edge of regression at the focus, the differential of $\mathbf{R}$ is parallel to $\mathbf{d}$. That is to say

$$(\mathbf{r}_1 du + \mathbf{r}_2 dv) + d\rho\,\mathbf{d} + \rho\,(\mathbf{d}_1 du + \mathbf{d}_2 dv)$$

is parallel to $\mathbf{d}$ and therefore perpendicular to $\mathbf{d_1}$ and $\mathbf{d_2}$. Hence, on forming its scalar product with $\mathbf{d_1}$ and $\mathbf{d_2}$ in turn, we have

$$\left.\begin{array}{l} (a\,du + b\,dv) + \rho\,(e\,du + f\,dv) = 0 \\ (b'\,du + c\,dv) + \rho\,(f\,du + g\,dv) = 0 \end{array}\right\} \quad \dots\dots\dots\dots(13).$$

These two equations determine ρ and du/dv. On eliminating ρ we have

$$\left| \begin{array}{cc} a\,du + b\,dv & b'\,du + c\,dv \\ e\,du + f\,dv & f\,du + g\,dv \end{array} \right| = 0 \dots\dots\dots\dots\dots(14),$$

which is a quadratic in du/dv, giving two directions, real or imaginary, for which Σ is a developable surface. Thus through each ray of the congruence there pass two developable surfaces, each with its edge of regression. Each ray of the congruence is therefore tangent to two curves in space, the points of contact being the *foci* or *focal points* of the ray. The locus of the foci of the rays is called the *focal surface* of the congruence. It is touched by all the rays of the congruence. The *focal planes* are the planes through the ray and the consecutive generators of the developable surfaces through the ray. They are the tangent planes at the foci to the two sheets of the focal surface.

On eliminating du/dv from the equations (13) we have

$$\left| \begin{array}{cc} a + e\rho & b + f\rho \\ b' + f\rho & c + g\rho \end{array} \right| = 0,$$

or $h^2\rho^2 + [ag - (b + b')f + ce]\,\rho + (ac - bb') = 0 \quad \dots(15),$

a quadratic in ρ giving the distances of the two foci from the director surface. It will be observed that this differs from the quadratic (7) only in the absolute term. Denoting the roots of (15) by ρ_1 and ρ_2 we have

$$\left.\begin{array}{l} h^2\,(\rho_1 + \rho_2) = (b + b')f - ec - ga \\ h^2\rho_1\rho_2 = ac - bb' \end{array}\right\} \quad \dots\dots\dots\dots(16).$$

Comparing these with the equations (8) for the limits we see that

$$\left.\begin{array}{l} r_1 + r_2 = \rho_1 + \rho_2 \\ (r_1 - r_2)^2 - (\rho_1 - \rho_2)^2 = \dfrac{(b - b')^2}{h^2} \end{array}\right\} \quad \dots\dots\dots\dots(17).$$

and

From the first of these relations it follows that the point midway between the limits is also midway between the foci. It is called the *middle point* of the ray ; and the locus of the middle points of all the rays is the *middle surface* of the congruence. From the second of equations (17) we see that, since the second member is positive, the distance between the foci is never greater than the distance between the limits. The two distances are equal only when $b = b'$. Thus on each ray there are five special points, the two limits, the two foci and the middle point. The foci, when they are real, lie between the limits.

We have seen that the central plane, P (through a given ray and the common perpendicular to it and a consecutive ray), is inclined to one of the principal planes at an angle θ given by

$$r = r_1 \cos^2 \theta + r_2 \sin^2 \theta.$$

The angle θ varies from 0 to $\dfrac{\pi}{2}$ as r varies from r_1 to r_2, the principal planes being perpendicular to each other. When the foot of the common perpendicular is one of the foci, the plane P is a focal plane. At the foci r has the values ρ_1 and ρ_2. Let θ_1 and θ_2 be the corresponding values of θ. Then

$$\left.\begin{aligned}\rho_1 &= r_1 \cos^2 \theta_1 + r_2 \sin^2 \theta_1 \\ \rho_2 &= r_1 \cos^2 \theta_2 + r_2 \sin^2 \theta_2\end{aligned}\right\}.$$

Adding these, and remembering that $\rho_1 + \rho_2 = r_1 + r_2$, we see that

$$\cos^2 \theta_1 + \cos^2 \theta_2 = 1,$$

and therefore, as θ_1 and θ_2 are both positive and neither is greater than $\dfrac{\pi}{2}$, we must have

$$\theta_1 + \theta_2 = \frac{\pi}{2} \quad\quad\quad\quad\quad\quad\text{(18).}$$

Thus the focal planes are not perpendicular, but are symmetrically placed with respect to the principal planes, so that *the planes bisecting the angles between the focal planes also bisect the angles between the principal planes.*

Further, the angle ϕ between the focal planes is given by

$$\phi = \theta_2 - \theta_1 = \frac{\pi}{2} - 2\theta_1.$$

Hence $\quad\quad \sin \phi = \cos 2\theta_1 = \cos^2 \theta_1 - \cos^2 \theta_2$

$$= \frac{\rho_1 - \rho_2}{r_1 - r_2} \quad\quad\quad\quad\quad\quad\text{(19).}$$

*98. Parameter of distribution. Consider again the con-secutive rays $\mathbf{d}$ and $\mathbf{d} + d\mathbf{d}$ corresponding to the parameter values u, v and $u + du, v + dv$. Then in virtue of the results proved in Art. 68 we see that, if

$$D = [\mathbf{r}', \mathbf{d}', \mathbf{d}],$$

the mutual moment of the two rays is $D ds^2$, their shortest distance apart is $D ds^2/d\sigma$, while the parameter of distribution for the ray $\mathbf{d}$ on the ruled surface determined by du/dv is

$$\beta = \frac{D}{\mathbf{d}'^2} = \frac{D ds^2}{d\sigma^2} \quad \ldots\ldots\ldots\ldots\ldots\ldots(20).$$

Now
$$D ds^2 = [d\mathbf{r}, d\mathbf{d}, \mathbf{d}]$$
$$= [\mathbf{r}_1 du + \mathbf{r}_2 dv, \ \mathbf{d}_1 du + \mathbf{d}_2 dv, \ \mathbf{d}].$$

Expanding this triple product according to the distributive law we see that the coefficient of du^2 is equal to

$$[\mathbf{r}_1, \mathbf{d}_1, \mathbf{d}] = \mathbf{r}_1 \cdot \mathbf{d}_1 \times \frac{\mathbf{d}_1 \times \mathbf{d}_2}{h}$$

$$= \frac{1}{h} \mathbf{r}_1 \cdot (f\mathbf{d}_1 - e\mathbf{d}_2)$$

$$= \frac{1}{h} (af - b'e).$$

Similarly the coefficient of $du\,dv$ is equal to

$$\frac{1}{h} \mathbf{r}_1 \cdot \mathbf{d}_2 \times (\mathbf{d}_1 \times \mathbf{d}_2) + \frac{1}{h} \mathbf{r}_2 \cdot \mathbf{d}_1 \times (\mathbf{d}_1 \times \mathbf{d}_2)$$

$$= \frac{1}{h} \mathbf{r}_1 \cdot (g\mathbf{d}_1 - f\mathbf{d}_2) + \frac{1}{h} \mathbf{r}_2 \cdot (f\mathbf{d}_1 - e\mathbf{d}_2)$$

$$= \frac{1}{h} (ag + bf - b'f - ce),$$

and the coefficient of dv^2 reduces to

$$\frac{1}{h} (bg - cf).$$

We may write the result in determinantal form as

$$D ds^2 = \frac{1}{h} \begin{vmatrix} a\,du + b\,dv, & b'\,du + c\,dv \\ e\,du + f\,dv, & f\,du + g\,dv \end{vmatrix},$$

so that the parameter of distribution has the value

$$\beta = \frac{\begin{vmatrix} a\,du + b\,dv, & b'\,du + c\,dv \\ e\,du + f\,dv, & f\,du + g\,dv \end{vmatrix}}{h\,(e\,du^2 + 2f\,du\,dv + g\,dv^2)} \quad \ldots\ldots\ldots\ldots(21).$$

For the developable surfaces of the congruence β vanishes identically. Equating this value of β to zero we have the same differential equation (14) for the developable surfaces of the congruence, found above by a different method.

*99. Mean ruled surfaces. The value of β as given by (21) is a function of du/dv: and therefore, for any one ray, the parameter of distribution is different for different surfaces through the ray. Those surfaces for which it has its greatest and least values are called the *mean surfaces* of the congruence through that ray. The differential equation of the mean ruled surfaces is obtained by equating to zero the derivative of β with respect to du/dv.

The analysis can be simplified by a suitable choice of the surface of reference and the parametric curves. Let the middle surface of the congruence be taken as director surface. Then, by (8) or (16), it follows that

$$f(b + b') - ag - ce = 0.$$

Further, let the parameters be chosen, as in Art. 96, so that the principal surfaces correspond to the parametric curves. Then, in virtue of (9), $f = b + b' = 0$. Thus our choice of surface and curves of reference gives the simplification

$$f = 0, \quad b + b' = 0, \quad ag + ce = 0 \quad \dots\dots\dots\dots(22).$$

The value of β as given by (21) then reduces to

$$\beta = \frac{be\,du^2 + 2ag\,du\,dv + bg\,dv^2}{\sqrt{eg}\,(e\,du^2 + g\,dv^2)}$$

$$= \frac{b}{\sqrt{eg}} + 2a\sqrt{\frac{g}{e}}\left(\frac{du\,dv}{e\,du^2 + g\,dv^2}\right)\dots\dots\dots\dots(23).$$

The values of du/dv corresponding to the stationary values of β are found by equating to zero the derivative of this expression with respect to du/dv. This leads to the equation

$$e\,du^2 - g\,dv^2 \quad \dots\dots\dots\dots\dots\dots\dots(24),$$

or, in virtue of (22), to

$$a\,du^2 + c\,dv^2 = 0 \quad \dots\dots\dots\dots\dots(24'),$$

as the differential equation of the mean surfaces. There are thus two mean surfaces through each ray. Now on the spherical representation the equation (24) is that of the curves bisecting the angles between the parametric curves, which correspond to the principal surfaces. Hence *the central planes for the mean surfaces*

w   13 *bisect the angles between the principal planes,* and therefore also the angles between the focal planes. Further, it follows from (5) and (24′) that the distance r to the central point of the ray, relative to a mean surface, is zero. Thus *the central point of a ray, relative to either of the mean surfaces, coincides with its middle point.* Both these results illustrate the appropriateness of the term "mean" as applied to these surfaces.

The extreme values of the parameter of distribution, corresponding to the mean surfaces, are obtained by substituting in (23) the values of du/dv given by (24). Denoting these values of β by β_1 and β_2 we have

$$\left.\begin{aligned}\beta_1 &= \frac{b}{\sqrt{eg}} + \frac{a}{e} \\[2mm] \beta_2 &= \frac{b}{\sqrt{eg}} - \frac{a}{e}\end{aligned}\right\}\ \dots\dots\dots\dots\dots(25).$$

The values of β for the principal surfaces are found from (23) by putting $dv = 0$ and $du = 0$ in turn. The two values so obtained are equal. Denoting them by $\bar{\beta}$ we have

$$\bar{\beta} = \frac{b}{\sqrt{eg}} = \tfrac{1}{2}(\beta_1 + \beta_2).$$

For this reason the parameter of distribution for a principal surface is called the *mean parameter* of the ray. It is the arithmetic mean of the extreme values of the parameter of distribution.

Let ϕ be the inclination of the central plane of a ray for any ruled surface, to the central plane for the mean surface $\sqrt{e}\,du = \sqrt{g}\,dv$, for which the parameter of distribution is β_1. We proceed to prove the formula

$$\beta = \beta_1 \cos^2 \phi + \beta_2 \sin^2 \phi \ \dots\dots\dots\dots\dots(26),$$

for the parameter of the ray relative to the first surface. The formula is analogous to (12), and is proved in a similar manner. The unit common perpendicular to consecutive rays of the first surface is given by (10), being

$$\frac{1}{\sqrt{eg}\,d\sigma}\,(g\mathbf{d}_1\,dv - e\mathbf{d}_2\,du).$$

For the mean surface $\sqrt{e}\,du = \sqrt{g}\,dv$ this becomes

$$\frac{1}{\sqrt{2eg}}(\sqrt{g}\,\mathbf{d}_1 - \sqrt{e}\,\mathbf{d}_2).$$

Hence the angle ϕ between the central planes for the two surfaces, being the angle between these two perpendiculars, is given by equating its cosine to the scalar product of the unit vectors. Thus

$$\cos \phi = \frac{\sqrt{e}\,du + \sqrt{g}\,dv}{\sqrt{2\,(e\,du^2 + g\,dv^2)}}.$$

Hence
$$\cos^2 \phi = \frac{1}{2} + \frac{\sqrt{eg}\,du\,dv}{e\,du^2 + g\,dv^2}$$

and therefore
$$\sin^2 \phi = \frac{1}{2} - \frac{\sqrt{eg}\,du\,dv}{e\,du^2 + g\,dv^2}.$$

Then, in virtue of (25),

$$\beta_1 \cos^2 \phi + \beta_2 \sin^2 \phi = \frac{b}{\sqrt{eg}} + 2a\sqrt{\frac{g}{e}}\left(\frac{du\,dv}{e\,du^2 + g\,dv^2}\right)$$
$$= \beta,$$

as required. If β_1 and β_2 have the same sign, it follows from this formula that the parameter of distribution has the same sign for all surfaces through the ray. Such rays are said to be *elliptic*. If, however, β_1 and β_2 have opposite signs, the parameter of distribution is positive for some surfaces, negative for others. Such rays are said to be *hyperbolic*. If either β_1 or β_2 is zero the ray is said to be *parabolic*.

For the developable surfaces of the congruence β vanishes identically. Hence the inclinations of the focal planes to the central plane for the mean surface corresponding to β_1 are given by

$$\tan \phi = \pm \sqrt{-\frac{\beta_1}{\beta_2}} \quad ....(27).$$

It follows, as already proved, that the central planes for the mean surfaces bisect the angles between the focal planes. If $\beta_1 = 0$ the ray is parabolic, and the focal planes coincide with the central plane for this mean surface. The two developables through the ray then coincide, and the foci coincide with the middle point of the ray. The two sheets of the focal surface are then identical. When this property holds for all the rays the congruence is said to be *parabolic*.

Ex. 1. Show that, for a principal surface, $\cos^2 \phi = \frac{1}{2}$. Deduce that the central planes for the mean surfaces bisect the angles between the principal surfaces.

Ex. 2. The foci are imaginary when the ray is elliptic.

100. Normal congruence. A congruence of straight lines is said to be *normal* when its rays are capable of orthogonal intersection by a surface, and therefore in general by a family of surfaces. Normal congruences were the first to be studied, especially in connection with the effects of reflection and refraction of rays of light. If this normal property is possessed by the congruence

$$\mathbf{R} = \mathbf{r} + t\mathbf{d},$$

there must be variations of $\mathbf{R}$ representing displacements perpendicular to $\mathbf{d}$, so that $\mathbf{d} \cdot d\mathbf{R} = 0$, that is

$$\mathbf{d} \cdot (d\mathbf{r} + \mathbf{d}\, dt + t\, d\mathbf{d}) = 0,$$

or $$\mathbf{d} \cdot d\mathbf{r} = -\, dt.$$

It follows that $\mathbf{d} \cdot d\mathbf{r}$ is a perfect differential. Also the analysis is reversible, and therefore we have *Hamilton's theorem* that *the necessary and sufficient condition that a congruence be normal is that* $\mathbf{d} \cdot d\mathbf{r}$ *be a perfect differential.* We may write

$$\mathbf{d} \cdot d\mathbf{r} = \mathbf{d} \cdot \mathbf{r}_1 du + \mathbf{d} \cdot \mathbf{r}_2 dv,$$

and if this is a perfect differential it follows that

$$\frac{\partial}{\partial v}(\mathbf{d} \cdot \mathbf{r}_1) = \frac{\partial}{\partial u}(\mathbf{d} \cdot \mathbf{r}_2),$$

or $$\mathbf{d}_2 \cdot \mathbf{r}_1 = \mathbf{d}_1 \cdot \mathbf{r}_2.$$

that is $$b = b'\dots\dots\dots\dots\dots\dots\dots\dots(28).$$

Conversely, if this relation holds, the congruence is normal. Further, if $b = b'$ it follows from (17) that

$$\rho_1 - \rho_2 = r_1 - r_2.$$

But $$\rho_1 + \rho_2 = r_1 + r_2,$$

and therefore $\rho_1 = r_1$ and $\rho_2 = r_2$. Hence *the foci coincide with the limits.* Also the focal planes coincide with the principal planes, and are therefore perpendicular to each other.

The assemblage of normals to a surface S has already been cited as an example of a normal congruence. The foci for any normal are the centres of curvature, and the focal surface is the centro-surface of S. The focal planes coincide with the principal planes, and are the principal normal planes of the surface S. The normals to S are also normals to any surface parallel to S. This agrees with the fact that when we integrate $dt = -\, \mathbf{d} \cdot d\mathbf{r}$, the second member being a perfect differential, the result involves an additive constant, which is arbitrary. Thus a normal congruence is cut orthogonally by a family of surfaces.

Ex. 1. For rays of a normal congruence $\beta_1 = -\beta_2$ and $\bar{\beta} = 0$.

Ex. 2. *The tangents to a singly infinite family of geodesics on a surface constitute a normal congruence* (Art. 81).

Ex. 3. The congruences considered in the examples of Art. 95 are normal.

Ex. 4. For the congruence of normals to a given surface, take the surface itself as director surface and its lines of curvature as parametric curves ($F = M = 0$). Show that, since $\mathbf{d} = \mathbf{n}$,

$$a = -L, \quad b = b' = 0, \quad c = -N,$$
$$e = JL - KE, \quad f = 0, \quad g = JN - KG,$$

and deduce that the equation for the focal distances from the surface is the equation for the principal radii of curvature.

101. Theorem of Malus and Dupin. *If a system of rays constituting a normal congruence is subjected to any number of reflections and refractions at the surfaces of successive homogeneous media, the congruence remains normal throughout.*

Consider the effect of reflection or refraction at the surface bounding two homogeneous media. Take this as director surface. Let $\mathbf{r}$ be the point of incidence of the ray whose initial direction is that of the unit vector $\mathbf{d}$, and which emerges parallel to $\mathbf{d}'$. Also let $\mathbf{n}$ be the unit normal to the surface at the point of incidence. Then, since the incident and refracted (or reflected) rays are coplanar with the normal, we may write

$$\mathbf{d} = \lambda \mathbf{n} + \mu \mathbf{d}'.$$

Hence $\mathbf{d} \times \mathbf{n} = \mu \mathbf{d}' \times \mathbf{n}.$

But, by the laws of reflection and refraction of light, $\mathbf{d} \times \mathbf{n}/\mathbf{d}' \times \mathbf{n}$ is constant for the same two media, being equal to the index of refraction in the case of refraction, or to unity in the case of reflection. Hence μ is constant for the congruence. Now, for a displacement $d\mathbf{r}$ along the surface, $\mathbf{n} \cdot d\mathbf{r} = 0$ and therefore

$$\mathbf{d} \cdot d\mathbf{r} = \mu \mathbf{d}' \cdot d\mathbf{r}.$$

But the first member of this equation is a perfect differential, since the incident system is a normal congruence. Consequently, μ being constant, $\mathbf{d}' \cdot d\mathbf{r}$ is also a perfect differential, and the emerging system likewise is a normal congruence. Thus the system remains normal after each reflection or refraction, and the theorem is proved.

*102. Isotropic congruence. When the coefficients of the two quadratic forms

$$a\,du^2 + (b + b')\,du\,dv + c\,dv^2 \quad \text{and} \quad e\,du^2 + 2f\,du\,dv + g\,dv^2$$

are proportional, that is to say when

$$a:b+b':c = e:2f:g \quad \dots\dots\dots\dots\dots(29),$$

the congruence is said to be *isotropic*. When these relations hold it follows from (5) that the central point of a ray is the same for all surfaces of the congruence which pass through the ray. Thus the two limits coincide with each other and with the middle point of the ray: and the limit surfaces coincide with each other and with the middle surface. The line of striction for any ruled surface of the congruence is the locus of the central points of its generators, and therefore the locus of the middle points of the rays. Thus the lines of striction of all surfaces of the congruence lie on the middle surface.

Let the middle surface be taken as director surface. Then the value of r as given by (5) must vanish identically, so that

$$a = 0, \quad b + b' = 0, \quad c = 0 \quad \dots\dots\dots\dots(30).$$

Hence
$$d\mathbf{r} \cdot d\mathbf{d} = 0 \quad \dots\dots\dots\dots\dots(30')$$

for any displacement $d\mathbf{r}$ on the middle surface. Thus if we make a spherical representation of the middle surface, making the point of that surface which is cut by the ray $\mathbf{d}$ correspond to the point $\mathbf{d}$ on the unit sphere, *any element of arc on the middle surface is perpendicular to the corresponding element on the spherical representation.*

Moreover, in virtue of the relations (29) it follows that the value of β as given by (21) is independent of the ratio $du:dv$. Hence *in an isotropic congruence the parameter of distribution for any ray has the same value for all surfaces through that ray.* Now, by (31) of Art. 71, the tangent plane to a ruled surface at a point of the generator distant u from the central point, is inclined to the central plane at an angle ϕ given by

$$\tan \phi = \frac{u}{\beta}$$

Hence, since the central point of a ray is the same for all surfaces through it, *any two surfaces of the congruence through a given ray cut each other along this ray at a constant angle.*

Ex. 1. On a ray of an isotropic congruence two points P, Q are taken at a given constant distance from the middle surface in opposite directions along the ray. Show that the surface generated by P is applicable to that generated by Q.

The points P and Q have position vectors $\mathbf{r}+t\mathbf{d}$ and $\mathbf{r}-t\mathbf{d}$, where t is constant. And for the locus of P

$$ds^2 = (d\mathbf{r}+t\,d\mathbf{d})^2 = (d\mathbf{r})^2 + t^2\,(d\mathbf{d})^2,$$

in virtue of (30'). The same result is obtained for the locus of Q.

Ex. 2. Deduce from (27) that, *for rays of an isotropic congruence, the foci and focal planes are imaginary.*

Ex. 3. The only normal isotropic congruence is a system of rays through a point.

Curvilinear Congruences

103. Congruence of curves. We shall now consider briefly the properties of a *curvilinear congruence*, which is a family of curves whose equations involve two independent parameters. If the curves are given as the lines of intersection of two families of surfaces, their equations are of the form

$$f(x, y, z, u, v) = 0, \quad g(x, y, z, u, v) = 0 \quad \ldots\ldots\ldots(31),$$

in which u, v are the parameters. In general only a finite number of curves will pass through a point (x_0, y_0, z_0). These are determined by the values of u, v which satisfy the equations

$$f(x_0, y_0, z_0, u, v) = 0, \quad g(x_0, y_0, z_0, u, v) = 0.$$

The curve which corresponds to the values u, v of the parameters is given by

$$f(u, v) = 0, \quad g(u, v) = 0 \quad \ldots\ldots\ldots\ldots\ldots(32).$$

A consecutive curve is given by

$$f(u + du, \ v + dv) = 0, \quad g(u + du, \ v + dv) = 0,$$

or by

$$\left.\begin{array}{l} f(u, v) + f_1 du + f_2 dv = 0 \\ g(u, v) + g_1 du + g_2 dv = 0 \end{array}\right\} \quad \ldots\ldots\ldots\ldots(33),$$

each value of the ratio du/dv determining a different consecutive curve. The curves (32) and (33) will intersect if the equations

$$f_1 du + f_2 dv = 0, \quad g_1 du + g_2 dv = 0$$

hold simultaneously, that is if

$$\frac{f_1}{g_1} = \frac{f_2}{g_2} = \lambda, \text{ (say)} \quad \dots \dots \dots \dots (34).$$

The points of intersection of the curve u, v with consecutive curves are given by (32) and (34). These points are called the *foci* or *focal points* of the curve. The locus of the foci of all the curves is the *focal surface* of the congruence. Its equation is obtained by eliminating u, v from the equations (32) and (34). The focal surface consists of as many sheets as there are foci on each curve.

The number of foci on any curve depends upon the nature of the congruence. The foci are found from the equations

$$f = 0, \quad g = 0, \quad f_1 g_2 - f_2 g_1 = 0 \quad \dots \dots \dots \dots (35).$$

In the case of a rectilinear congruence, $f = 0$ and $g = 0$ are planes. Hence the above equations are of order 1, 1, 2 respectively in the coordinates, showing that there are two foci (real or imaginary) on any ray. The two may of course coincide as in a parabolic congruence. In the case of a congruence of conics we may suppose that $f = 0$ is a conicoid and $g = 0$ a plane. The equations determining the foci are then of order 2, 1, 3 respectively in the coordinates, and there are six foci on each conic.

104. Surfaces of the congruence. As in the case of a rectilinear congruence, the various curves may be grouped so as to constitute surfaces. Any assumed relation between the parameters determines such a surface. Taking the relation

$$v = \phi(u)$$

and eliminating u, v between this equation and the equations of the congruence, we obtain a relation between x, y, z, representing one of the *surfaces of the congruence*. Each relation between the parameters gives such a surface. There is thus an infinitude of surfaces corresponding to the infinitude of forms for the relation between the parameters.

We shall now prove the theorem that *all the surfaces of the congruence, which pass through a given curve, touch one another as well as the focal surface at the foci of that curve.* Consider the surface through the curve u, v determined by the equations

$$f = 0, \quad g = 0, \quad v = \phi(u) \quad \dots \dots \dots \dots (36).$$

Then any displacement (dx, dy, dz) on that surface is such that

$$\left. \begin{aligned} \frac{\partial f}{\partial x} dx + \frac{\partial f}{\partial y} dy + \frac{\partial f}{\partial z} dz + \left[\frac{\partial f}{\partial u} + \frac{\partial f}{\partial v} \phi'(u) \right] du = 0 \\ \frac{\partial g}{\partial x} dx + \frac{\partial g}{\partial y} dy + \frac{\partial g}{\partial z} dz + \left[\frac{\partial g}{\partial u} + \frac{\partial g}{\partial v} \phi'(u) \right] du = 0 \end{aligned} \right\} .$$

But at the foci

$$\frac{f_1}{g_1} = \frac{f_2}{g_2} = \lambda,$$

and therefore for any direction in the tangent plane at a focus we must have

$$\frac{\partial f}{\partial x} dx + \frac{\partial f}{\partial y} dy + \frac{\partial f}{\partial z} dz = \lambda \left(\frac{\partial g}{\partial x} dx + \frac{\partial g}{\partial y} dy + \frac{\partial g}{\partial z} dz \right).$$

Thus the normal to the surface at the focus is parallel to the vector

$$\left(\frac{\partial f}{\partial x} - \lambda \frac{\partial g}{\partial x}, \quad \frac{\partial f}{\partial y} - \lambda \frac{\partial g}{\partial y}, \quad \frac{\partial f}{\partial z} - \lambda \frac{\partial g}{\partial z} \right) \quad \dots\dots\dots(37),$$

which is independent of the assumed relation between the parameters, and is therefore the same for all surfaces of the congruence through the given curve. Hence all these surfaces touch one another at the foci of the curve.

Again, the equation of the focal surface is the eliminant of u, v from the equations (32) and (34). At any point of the focal surface we have from the first of these

$$\left. \begin{aligned} \frac{\partial f}{\partial x} dx + \frac{\partial f}{\partial y} dy + \frac{\partial f}{\partial z} dz + \frac{\partial f}{\partial u} du + \frac{\partial f}{\partial v} dv = 0 \\ \frac{\partial g}{\partial x} dx + \frac{\partial g}{\partial y} dy + \frac{\partial g}{\partial z} dz + \frac{\partial g}{\partial u} du + \frac{\partial g}{\partial v} dv = 0 \end{aligned} \right\} ,$$

and therefore, in virtue of (34),

$$\frac{\dfrac{\partial f}{\partial x} dx + \dfrac{\partial f}{\partial y} dy + \dfrac{\partial f}{\partial z} dz}{\dfrac{\partial g}{\partial x} dx + \dfrac{\partial g}{\partial y} dy + \dfrac{\partial g}{\partial z} dz} = \frac{f_1}{g_1} = \frac{f_2}{g_2} = \lambda.$$

Thus (dx, dy, dz) is perpendicular to the vector (37), which is therefore normal to the focal surface. Hence at a focus of the curve the focal surface has the same tangent plane as any surface of the congruence which passes through the curve. The theorem is thus established. It follows that *any surface of the congruence touches the focal surface at the foci of all its curves.* The tangent planes to the focal surface at the foci of a curve are called the *focal planes* of the curve.

105. Normal congruence. A curvilinear congruence is said to be *normal* when it is capable of orthogonal intersection by a family of surfaces. Let the congruence be given by the equations (31). Along a particular curve the parameters u, v are constant, and therefore, for a displacement along the curve, we have on differentiation

$$\frac{dx}{J\left(\frac{f,g}{y,z}\right)} = \frac{dy}{J\left(\frac{f,g}{z,x}\right)} = \frac{dz}{J\left(\frac{f,g}{x,y}\right)}.$$

If in this equation we substitute the values of u, v in terms of x, y, z as given by (31), we obtain the differential equation of the curves of the congruence in the form

$$\frac{dx}{X} = \frac{dy}{Y} = \frac{dz}{Z} \quad\dots\dots\dots\dots\dots\dots\dots(38),$$

where X, Y, Z are independent of u, v. If then the congruence is normal to a surface, the differential equation of the surface must be

$$X\,dx + Y\,dy + Z\,dz = 0 \quad\dots\dots\dots\dots\dots(39).$$

In general this equation is not integrable. It is well known from the theory of differential equations that the condition of integrability is

$$X\left(\frac{\partial Y}{\partial z} - \frac{\partial Z}{\partial y}\right) + Y\left(\frac{\partial Z}{\partial x} - \frac{\partial X}{\partial z}\right) + Z\left(\frac{\partial X}{\partial y} - \frac{\partial Y}{\partial x}\right) = 0.$$

If this condition is satisfied there is a family of surfaces satisfying the equation (39), and therefore cutting the congruence orthogonally.

Ex. 1. The congruence of circles

$$lx + my + nz = u, \quad x^2 + y^2 + z^2 = v$$

has for its differential equation

$$\frac{dx}{ny - mz} = \frac{dy}{lz - nx} = \frac{dz}{mx - ly}.$$

Hence they are normal to the surfaces given by

$$(ny - mz)\,dx + (lz - nx)\,dy + (mx - ly)\,dz = 0.$$

The condition of integrability is satisfied, and the integral may be expressed

$$ny - mz = c(nx - lz),$$

where c is an arbitrary constant. This represents a family of planes with common line of intersection $x/l = y/m = z/n$.

Ex. 2. The congruence of conics

$$y - z = u, \quad (y+z)^2 - 4x = v$$

has a differential equation

$$\frac{dx}{y+z} = dy = dz.$$

It is normal to the surfaces given by

$$(y+z)\,dx + dy + dz = 0.$$

The condition of integrability is satisfied, and the integral is

$$y + z = ce^{-x}.$$

Ex. 3. Show that the congruence of circles

$$x^2 + y^2 + z^2 = uy = vz$$

has the differential equation

$$\frac{dx}{x^2 - y^2 - z^2} = \frac{dy}{2xy} = \frac{dz}{2xz},$$

and is cut orthogonally by the family of spheres

$$x^2 + y^2 + z^2 = cx.$$

EXAMPLES XIII

Rectilinear Congruences

1. The current point on the middle surface is

$$\mathbf{R} = \mathbf{r} + t\mathbf{d},$$

where
$$t = \frac{1}{2h^2}\left[f(b+b') - ec - ga \right].$$

The condition that the surface of reference may be the middle surface is

$$ec + ga = f(b+b').$$

2. Prove that, on each sheet of the focal surface, the curves corresponding to the two families of developable surfaces of the congruence are conjugate.

3. The tangent planes to two confocal quadrics at the points of contact of a common tangent are perpendicular. Hence show that the common tangents to two confocal quadrics form a normal congruence.

4. If two surfaces of a congruence through a given ray are represented on the unit sphere by curves which cut orthogonally, their lines of striction meet the ray at points equidistant from the middle point.

5. Through each point of the plane $z=0$ a ray (l, m, n) is drawn, such that

$$l = ky, \quad m = -kx, \quad n = \sqrt{1 - k^2(x^2 + y^2)}.$$

Show that the congruence so formed is isotropic, with the plane $z=0$ as middle surface.

6. In Ex. 1, Art. 102, prove conversely that, if the surfaces are applicable, the congruence is isotropic.

7. If (l, m, n) is the unit normal to a minimal surface at the current point (x, y, z), the line parallel to $(m, -l, n)$ through the point $(x, y, 0)$ generates a normal congruence.

8. The lines of striction of the mean ruled surfaces lie on the middle surface.

9. For any choice of parameters the differential equation of the mean surfaces of a congruence is

$$\frac{ag-(b+b')f+ce}{2h^2} = \frac{a\,du^2+(b+b')\,du\,dv+c\,dv^2}{e\,du^2+2f\,du\,dv+g\,dv^2}.$$

10. The mean parameter of distribution (Art. 99) of a congruence is the square root of the difference of the squares of the distances between the limits and between the foci.

11. If the two sheets of the focal surface intersect, the curve of intersection is the envelope of the edges of regression of the two families of developable surfaces of the congruence.

12. In the congruence of straight lines which intersect two twisted curves, whose arc-lengths are s, s', the differential equation of the developable surfaces of the congruence is $ds\,ds' = 0$. The focal planes for a ray are the planes through the ray and the tangents to the curves at the points where it cuts them.

13. One end of an inextensible thread is attached to a fixed point on a smooth surface, and the thread is pulled tightly over the surface. Show that the possible positions of its straight portions form a normal congruence, and that a particle of the thread describes a normal surface.

14. In the *congruence of tangents to one system of asymptotic lines on a given surface*, S, show that the two sheets of the focal surface coincide with each other and with the surface S, and that the distance between the limit points of a ray is equal to $1/\sqrt{-K}$, K being the specific curvature of the surface S at the point of contact of the ray.

Take the surface S as director surface, the given system of asymptotic lines as the parametric curves $v = \text{const.}$, and their orthogonal trajectories as the curves $u = \text{const.}$ Then, for the surface S,

$$L = 0, \quad F = 0,$$

so that

$$K = -\frac{M^2}{EG}.$$

Also, with the usual notation, $\mathbf{d} = \mathbf{r}_1/\sqrt{E}$ and therefore

$$\left.\begin{aligned}
\mathbf{d}_1 &= \frac{\mathbf{r}_{11}}{\sqrt{E}} + \mathbf{r}_1 \frac{\partial}{\partial u}\left(\frac{1}{\sqrt{E}}\right) = -\frac{E_2}{2G\sqrt{E}}\mathbf{r}_2 \\
\mathbf{d}_2 &= \frac{\mathbf{r}_{12}}{\sqrt{E}} + \mathbf{r}_1 \frac{\partial}{\partial v}\left(\frac{1}{\sqrt{E}}\right) = \frac{M}{\sqrt{E}}\mathbf{n} + \frac{G_1}{2G\sqrt{E}}\mathbf{r}_2
\end{aligned}\right\} \text{ by Art. 41.}$$

From these it is easily verified that

$$a=0, \quad b=-\frac{E_2}{2\sqrt{E}}, \quad b'=0, \quad c=\frac{G_1}{2\sqrt{E}},$$

$$e=\frac{E_2^2}{4EG}, \quad f=-\frac{E_2 G_1}{4EG}, \quad g=\frac{4M^2G+G_1^2}{4EG},$$

$$h^2=\frac{M^2 E_2^2}{4E^2G}.$$

The equation (15), for the distances of the foci from the director surface, reduces to $\rho^2=0$. Thus the foci coincide; and the two sheets of the focal surface coincide with the surface S. The congruence is therefore *parabolic* (Art. 99). Similarly the equation (7), for the distances of the limits, reduces to

$$r^2=\frac{EG}{4M^2}=\frac{1}{-4K},$$

or
$$r=\pm\frac{1}{2\sqrt{-K}}.$$

Thus the distance between the limits is $1/\sqrt{-K}$.

15. When the two sheets of the focal surface of a rectilinear congruence coincide, the specific curvature of the focal surface at the point of contact of a ray is $-1/l^2$, where l is the distance between the limits of the ray.

16. If, in a normal congruence, the distance between the foci of a ray is the same for all rays, show that the two sheets of the focal surface have their specific curvature constant and negative.

17. Rays are incident upon a reflecting surface, and the developables of the incident congruence are reflected into the developables of the reflected congruence. Show that they cut the reflecting surface in conjugate lines.

18. When a congruence consists of the tangents to one system of lines of curvature on a surface, the focal distances are equal to the radii of geodesic curvature of the other system of lines of curvature.

19. A necessary and sufficient condition that the tangents to a family of curves on a surface may form a normal congruence is that the curves be geodesics.

20. The extremities of a straight line, whose length is constant and whose direction depends upon two parameters, are made to describe two surfaces applicable to each other. Show that the positions of the line form an isotropic congruence.

21. The spherical representations (Art. 94) of the developable surfaces of an isotropic congruence are null lines.

22. In an isotropic congruence the envelope of the plane which cuts a ray orthogonally at its middle point is a minimal surface.

Curvilinear Congruences

23. Prove that the congruence

$$x - y = u(z - x) \quad \Big\}$$
$$(x - y)^2(x + y + z) = v \Big\}$$

is normal, being cut orthogonally by the family of surfaces

$$yz + zx + xy = c^2.$$

24. Show that

$$2x^2 - y^2 = u^2, \quad 3y^2 + 2z^2 = v^2$$

represents a normal congruence, cut orthogonally by the surfaces

$$xy^2 = cz^3.$$

25. Four surfaces of the congruence pass through a given curve of the congruence. Show that the cross-ratio of their four tangent planes at a point of the curve is independent of the point chosen.

26. If the curves of a congruence cut a fixed curve, C, each point of intersection is a focal point, unless the tangents at this point to all curves of the congruence which pass through it, are coplanar with the tangent to the curve C at the same point.

27. If all the curves of a congruence meet a fixed curve, this fixed curve lies on the focal surface.

28. Show that the congruence

$$a\phi(x) + b\psi(y) + c\chi(z) = u \Big\}$$
$$\phi(x) + \psi(y) + \chi(z) = v \Big\}$$

is normal to a family of surfaces; and determine the family.

29. Find the parallel plane sections of the surfaces

$$\phi(x) + \psi(y) + \chi(z) = u$$

which constitute a normal congruence; and determine the family of surfaces which cut them orthogonally.

30. If a congruence of circles is cut orthogonally by more than two surfaces, it is cut orthogonally by a family of surfaces. Such a congruence is called a *cyclic system*.

NOTE. The author has recently shown that curvilinear congruences may be more effectively treated along the same lines as rectilinear congruences. The existence of a limit surface and a surface of striction is thus easily established, and the equations of these surfaces are readily found. See Art. 129 below.

CHAPTER XI

TRIPLY ORTHOGONAL SYSTEMS OF SURFACES

106. A triply orthogonal system consists of three families of surfaces

$$\left.\begin{aligned} u\,(x,\,y,\,z) &= \text{const.} \\ v\,(x,\,y,\,z) &= \text{const.} \\ w\,(x,\,y,\,z) &= \text{const.} \end{aligned}\right\} \quad \ldots\ldots\ldots\ldots\ldots\ldots(1)$$

which are such that, through each point of space passes one and only one member of each family, each of the three surfaces cutting the other two orthogonally. The simplest example of such a system is afforded by the three families of planes

$$x = \text{const.}, \quad y = \text{const.}, \quad z = \text{const.},$$

parallel to the rectangular coordinate planes. Or again, if space is mapped out in terms of spherical polar coordinates r, θ, ϕ, the surfaces $r = \text{const.}$ are concentric spheres, the surfaces $\theta = \text{const.}$ are coaxial circular cones, and the surfaces $\phi = \text{const.}$ are the meridian planes. These three families form a triply orthogonal system. Another example is afforded by a family of parallel surfaces and the two families of developables in the congruence of normals (Arts. 74, 100). The developables are formed by the normals along the lines of curvature on any one of the parallel surfaces. As a last example may be mentioned the three families of quadrics confocal with the central quadric

$$\frac{x^2}{a} + \frac{y^2}{b} + \frac{z^2}{c} = 1.$$

It is well known that one of these is a family of ellipsoids, one a family of hyperboloids of one sheet, and the third a family of hyperboloids of two sheets. This example will soon be considered in further detail.

107. Normals. The values of u, v, w for the three surfaces through a point are called the *curvilinear coordinates* of the point. By means of the equations (1) the rectangular coordinates x, y, z, and therefore the position vector $\mathbf{r}$, of any point in space may be expressed in terms of the curvilinear coordinates. We assume that this has been done; and we denote partial derivatives with respect to u, v, w by the suffixes 1, 2, 3 respectively. Thus

$$\mathbf{r}_1 = \frac{\partial \mathbf{r}}{\partial u}, \quad \mathbf{r}_3 = \frac{\partial \mathbf{r}}{\partial w}, \quad \mathbf{r}_{12} = \frac{\partial^2 \mathbf{r}}{\partial u \partial v},$$

and so on.

The normal to the surface $u = $ const. at the point (x, y, z) is parallel to the vector

$$\left(\frac{\partial u}{\partial x}, \frac{\partial u}{\partial y}, \frac{\partial u}{\partial z} \right).$$

Let $\mathbf{a}$ denote the *unit* normal in the direction of u increasing. Similarly let $\mathbf{b}$ and $\mathbf{c}$ denote unit normals to the surfaces $v = $ const.

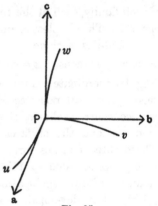

Fig. 25.

and $w = $ const. respectively, in the directions of v increasing and w increasing. Further we may take the three families in that cyclic order for which $\mathbf{a}$, $\mathbf{b}$, $\mathbf{c}$ are a right-handed system of unit vectors. Then since they are mutually perpendicular we have

$$\mathbf{a} \cdot \mathbf{b} = \mathbf{b} \cdot \mathbf{c} = \mathbf{c} \cdot \mathbf{a} = 0$$

and $$\mathbf{a} = \mathbf{b} \times \mathbf{c}, \quad \mathbf{b} = \mathbf{c} \times \mathbf{a}, \quad \mathbf{c} = \mathbf{a} \times \mathbf{b}$$ ....(2).

And, because they are unit vectors,

$$\mathbf{a}^2 = \mathbf{b}^2 = \mathbf{c}^2 = 1 \quad ....(3).$$

Since the normal to the surface $u = $ const. is tangential to the surfaces $v = $ const. and $w = $ const. through the point considered, for a displacement ds in the direction of $\mathbf{a}$ both v and w are constant. In terms of the change du in the other parameter let

$$ds = p\,du.$$

Thus $p\,du$ is the length of an element of arc normal to the surface $u = $ const. The unit normal in this direction is therefore given by

$$\mathbf{a} = \frac{d\mathbf{r}}{ds} = \frac{1}{p}\frac{\partial\mathbf{r}}{\partial u} = \frac{1}{p}\mathbf{r}_1,$$

so that $\qquad\qquad\qquad \mathbf{r}_1 = p\mathbf{a}$ ....(4),

and therefore $\qquad\qquad\quad \mathbf{r}_1{}^2 = p^2.$

Similarly if the elements of arc normal to the other two surfaces, in the directions of $\mathbf{b}$ and $\mathbf{c}$, are $q\,dv$ and $r\,dw$ respectively, we have

$$\mathbf{r}_2 = q\mathbf{b}, \quad \mathbf{r}_3 = r\mathbf{c} \ \text{........................(5)},$$

and consequently $\qquad \mathbf{r}_2{}^2 = q^2, \quad \mathbf{r}_3{}^2 = r^2.$

Thus $\mathbf{r}_1, \mathbf{r}_2, \mathbf{r}_3$ are a right-handed set of mutually perpendicular vectors, so that

$$\mathbf{r}_1 \boldsymbol{\cdot} \mathbf{r}_2 = \mathbf{r}_2 \boldsymbol{\cdot} \mathbf{r}_3 = \mathbf{r}_3 \boldsymbol{\cdot} \mathbf{r}_1 = 0 \ \text{...................(6)}.$$

Further, in virtue of (2), (4) and (5),

$$\left.\begin{array}{c} \mathbf{r}_2 \times \mathbf{r}_3 = \dfrac{qr}{p}\,\mathbf{r}_1 \\[2mm] \mathbf{r}_3 \times \mathbf{r}_1 = \dfrac{rp}{q}\,\mathbf{r}_2 \\[2mm] \mathbf{r}_1 \times \mathbf{r}_2 = \dfrac{pq}{r}\,\mathbf{r}_3 \end{array}\right\} \ \text{.......................(7)},$$

and $\qquad\qquad [\mathbf{r}_1, \mathbf{r}_2, \mathbf{r}_3] = pqr\,[\mathbf{a}, \mathbf{b}, \mathbf{c}] = pqr$ ....(8).

108. Fundamental magnitudes. A surface $u = $ const. is cut by those of the other two families in two families of curves, $v = $ const. and $w = $ const. Thus for points on a surface $u = $ const. we may take v, w as parametric variables. Similarly on a surface $v = $ const. the parameters are w, u and so on. Thus the parametric curves on any surface are its curves of intersection with members of the other families. On a surface $u = $ const. the fundamental magnitudes of the first order are therefore

$$\left.\begin{array}{l} E = \mathbf{r}_2{}^2 = q^2 \\ F = \mathbf{r}_2 \boldsymbol{\cdot} \mathbf{r}_3 = 0 \\ G = \mathbf{r}_3{}^2 = r^2 \end{array}\right\} \ \text{.......................(9)},$$

so that $\qquad\qquad\qquad H^2 = q^2 r^2,$

and similarly for the other surfaces. Since $F = 0$ the parametric curves on any surface constitute an orthogonal system.

To find the fundamental magnitudes of the second order we examine the second derivatives of $\mathbf{r}$. By differentiating the equations (6) with respect to w, u, v respectively, we have

$$\left. \begin{array}{c} \mathbf{r}_{13} \cdot \mathbf{r}_2 + \mathbf{r}_1 \cdot \mathbf{r}_{23} = 0 \\ \mathbf{r}_{21} \cdot \mathbf{r}_3 + \mathbf{r}_2 \cdot \mathbf{r}_{31} = 0 \\ \mathbf{r}_{32} \cdot \mathbf{r}_1 + \mathbf{r}_3 \cdot \mathbf{r}_{12} = 0 \end{array} \right\}.$$

Subtracting the second and third of these, and comparing the result with the first, we see immediately that

$$\left. \begin{array}{c} \mathbf{r}_1 \cdot \mathbf{r}_{23} = \mathbf{r}_2 \cdot \mathbf{r}_{31} = 0 \\ \mathbf{r}_3 \cdot \mathbf{r}_{12} = 0 \end{array} \right\} \quad \dots\dots\dots\dots\dots(10).$$

Similarly

Again, by differentiating $\mathbf{r}_1{}^2 = p^2$ with respect to u, v, w, we have

$$\left. \begin{array}{c} \mathbf{r}_1 \cdot \mathbf{r}_{11} = pp_1 \\ \mathbf{r}_1 \cdot \mathbf{r}_{12} = pp_2 \\ \mathbf{r}_1 \cdot \mathbf{r}_{13} = pp_3 \end{array} \right\} \quad \dots\dots\dots\dots\dots(11),$$

and therefore

$$\left. \begin{array}{c} \mathbf{r}_2 \cdot \mathbf{r}_{11} = -\mathbf{r}_1 \cdot \mathbf{r}_{12} = -pp_2 \\ \mathbf{r}_3 \cdot \mathbf{r}_{11} = -\mathbf{r}_1 \cdot \mathbf{r}_{13} = -pp_3 \end{array} \right\} \quad \dots\dots\dots(12),$$

with two similar sets of equations. Now the unit normal to the surface $u = \text{const.}$ is $\mathbf{r}_1/p$, and the parameters are v, w. Hence the second order magnitudes for that surface have the values

$$\left. \begin{array}{c} L = \dfrac{1}{p}\mathbf{r}_1 \cdot \mathbf{r}_{22} = -\dfrac{1}{p} q q_1 \\[2mm] M = \dfrac{1}{p}\mathbf{r}_1 \cdot \mathbf{r}_{23} = 0 \\[2mm] N = \dfrac{1}{p}\mathbf{r}_1 \cdot \mathbf{r}_{33} = -\dfrac{1}{p} r r_1 \end{array} \right\} \quad \dots\dots\dots\dots(13).$$

Similar results may be written down for the surfaces $v = \text{const.}$ and $w = \text{const.}$ They are collected for reference in the table*.

Surface	Parameters	E	F	G	H	L	M	N
$u = \text{const.}$	v, w	q^2	0	r^2	qr	$-\dfrac{1}{p} q q_1$	0	$-\dfrac{1}{p} r r_1$
$v = \text{const.}$	w, u	r^2	0	p^2	rp	$-\dfrac{1}{q} r r_2$	0	$-\dfrac{1}{q} p p_2$
$w = \text{const.}$	u, v	p^2	0	q^2	pq	$-\dfrac{1}{r} p p_3$	0	$-\dfrac{1}{r} q q_3$

* Forsyth gives a similar table on p. 413 of his "Lectures."

Ex. Elliptic coordinates. Consider the quadrics confocal with the ellipsoid

$$\frac{x^2}{a^2} + \frac{y^2}{b^2} + \frac{z^2}{c^2} = 1,$$

in which we may assume $a^2 > b^2 > c^2$. The confocals are given by

$$\frac{x^2}{a^2 + \lambda} + \frac{y^2}{b^2 + \lambda} + \frac{z^2}{c^2 + \lambda} = 1,$$

for different values of λ. Hence the values of λ for the confocals through a given point (x, y, z) are given by the cubic equation

$$\phi(\lambda) \equiv (a^2 + \lambda)(b^2 + \lambda)(c^2 + \lambda) - \Sigma x^2 (b^2 + \lambda)(c^2 + \lambda) = 0.$$

Let u, v, w denote the roots of this equation. Then, since the coefficient of λ^3 is equal to unity, we have

$$(\lambda - u)(\lambda - v)(\lambda - w) \equiv (a^2 + \lambda)(b^2 + \lambda)(c^2 + \lambda) - \Sigma x^2 (b^2 + \lambda)(c^2 + \lambda).$$

If in this identity we give λ the values $-a^2$, $-b^2$, $-c^2$ in succession, we find

$$\left. \begin{array}{l} x^2 = \dfrac{(a^2 + u)(a^2 + v)(a^2 + w)}{(a^2 - b^2)(a^2 - c^2)} \\[2ex] y^2 = \dfrac{(b^2 + u)(b^2 + v)(b^2 + w)}{(b^2 - c^2)(b^2 - a^2)} \\[2ex] z^2 = \dfrac{(c^2 + u)(c^2 + v)(c^2 + w)}{(c^2 - a^2)(c^2 - b^2)} \end{array} \right\} \quad \dots\dots\dots\dots\dots(14).$$

These equations give the Cartesian coordinates in terms of the parameters u, v, w, which are called the *elliptic coordinates* of the point (x, y, z).

By logarithmic differentiation of (14) we find

$$\mathbf{r}_1 = \left(\frac{\partial x}{\partial u}, \frac{\partial y}{\partial u}, \frac{\partial z}{\partial u} \right) = \frac{1}{2} \left(\frac{x}{a^2 + u}, \frac{y}{b^2 + u}, \frac{z}{c^2 + u} \right),$$

with similar expressions for $\mathbf{r}_2$ and $\mathbf{r}_3$. From these the relations (6) are easily verified; and further

$$p^2 = \mathbf{r}_1{}^2 = \frac{(u - v)(u - w)}{4(a^2 + u)(b^2 + u)(c^2 + u)},$$

$$q^2 = \mathbf{r}_2{}^2 = \frac{(v - w)(v - u)}{4(a^2 + v)(b^2 + v)(c^2 + v)},$$

$$r^2 = \mathbf{r}_3{}^2 = \frac{(w - u)(w - v)}{4(a^2 + w)(b^2 + w)(c^2 + w)}.$$

These are the first order magnitudes E, G for the confocal surfaces, F being equal to zero; and by partial differentiation we may calculate L, N according to the above table.

109. Dupin's theorem. We have seen that, for each surface of a triply orthogonal system, $F = 0$ and $M = 0$. Thus the parametric curves are lines of curvature, and we have Dupin's theorem:

The curves of intersection of the surfaces of a triply orthogonal system are lines of curvature on each.

The principal curvatures on each of the surfaces are then easily calculated. On a surface $u = $ const. let κ_{uv} denote the principal curvature in the direction of the curve of parameter v (the curve $w = $ const.), and κ_{uw} the principal curvature in the direction of the curve of parameter w (the curve $v = $ const.). Then

$$\kappa_{uv} = \frac{L}{E} = -\frac{q_1}{pq}$$
$$\kappa_{uw} = \frac{N}{G} = -\frac{r_1}{pr} \qquad \dots\dots\dots\dots\dots(15).$$

Similarly on a surface $v = $ const. the principal curvatures in the directions of the curves of parameters w and u are respectively

$$\kappa_{vw} = \frac{L}{E} = -\frac{r_2}{qr}$$
$$\kappa_{vu} = \frac{N}{G} = -\frac{p_2}{qp} \qquad \dots\dots\dots\dots\dots(16),$$

and on a surface $w = $ const. in the directions of the curves of parameters u, v they are respectively

$$\kappa_{wu} = \frac{L}{E} = -\frac{p_3}{rp}$$
$$\kappa_{wv} = \frac{N}{G} = -\frac{q_3}{rq} \qquad \dots\dots\dots\dots\dots(17).$$

Let κ_u be the curvature of the curve of parameter u. Then since κ_{wu} and κ_{vu} are the resolved parts of the vector curvature of this curve in the directions of the normals **c** and **b** respectively, we have (Art. 53)

$$\kappa_u \cos \varpi = \kappa_{wu}, \quad \kappa_u \sin \varpi = \kappa_{vu} \qquad \dots\dots\dots(18),$$

where ϖ is the normal angle of the curve relative to the surface $w = $ const. Hence

$$\kappa_u{}^2 = \kappa_{wu}{}^2 + \kappa_{vu}{}^2$$

and
$$\tan \varpi = \frac{\kappa_{vu}}{\kappa_{wu}} = \frac{rp_2}{qp_3},$$

with similar results for the curves of parameters v and w. Further, since the curve of parameter u is a line of curvature on the surface $w = $ const., the torsion W of its geodesic tangent is zero. Hence, by Art. 50, its own *torsion* τ is given by

$$\tau = -\frac{d\varpi}{ds} = -\frac{1}{p}\frac{\partial \varpi}{\partial u} \qquad \dots\dots\dots\dots(19).$$

110. Second derivatives of r. Explicit expressions for the second derivatives of **r** in terms of $\mathbf{r}_1$, $\mathbf{r}_2$, $\mathbf{r}_3$ are easily calculated. The resolved parts of $\mathbf{r}_{11}$ in the directions of the normals **a, b, c** are respectively

$$\mathbf{r}_{11} \cdot \mathbf{a}, \quad \mathbf{r}_{11} \cdot \mathbf{b}, \quad \mathbf{r}_{11} \cdot \mathbf{c},$$

or

$$\frac{1}{p}\mathbf{r}_1 \cdot \mathbf{r}_{11}, \quad \frac{1}{q}\mathbf{r}_2 \cdot \mathbf{r}_{11}, \quad \frac{1}{r}\mathbf{r}_3 \cdot \mathbf{r}_{11},$$

which, in virtue of (11) and (12), are equal to

$$p_1, \quad -\frac{1}{q}pp_2, \quad -\frac{1}{r}pp_3.$$

Hence we may write

$$
\left.
\begin{aligned}
\mathbf{r}_{11} &= \frac{1}{p}p_1\mathbf{r}_1 - \frac{p}{q^2}p_2\mathbf{r}_2 - \frac{p}{r^2}p_3\mathbf{r}_3 \\
\mathbf{r}_{22} &= \frac{1}{q}q_2\mathbf{r}_2 - \frac{q}{r^2}q_3\mathbf{r}_3 - \frac{q}{p^2}q_1\mathbf{r}_1 \\
\mathbf{r}_{33} &= \frac{1}{r}r_3\mathbf{r}_3 - \frac{r}{p^2}r_1\mathbf{r}_1 - \frac{r}{q^2}r_2\mathbf{r}_2
\end{aligned}
\right\} \quad \dots\dots\dots\dots(20).
$$

and similarly

In the same way we find that the resolved parts of $\mathbf{r}_{23}$ in the directions of **a, b, c** are respectively

$$0, \quad q_3, \quad r_2.$$

Hence the result

$$
\left.
\begin{aligned}
\mathbf{r}_{23} &= \frac{1}{q}q_3\mathbf{r}_2 + \frac{1}{r}r_2\mathbf{r}_3 \\
\mathbf{r}_{31} &= \frac{1}{r}r_1\mathbf{r}_3 + \frac{1}{p}p_3\mathbf{r}_1 \\
\mathbf{r}_{12} &= \frac{1}{p}p_2\mathbf{r}_1 + \frac{1}{q}q_1\mathbf{r}_2
\end{aligned}
\right\} \quad \dots\dots\dots\dots(21).
$$

and similarly

We may also calculate the derivatives of the unit normals **a, b, c.** For

$$\frac{\partial \mathbf{a}}{\partial u} = \frac{\partial}{\partial u}\left(\frac{\mathbf{r}_1}{p}\right) = \frac{1}{p}\mathbf{r}_{11} - \frac{1}{p^2}p_1\mathbf{r}_1$$

$$= -\frac{1}{q^2}p_2\mathbf{r}_2 - \frac{1}{r^2}p_3\mathbf{r}_3 \qquad \text{by (20)}$$

$$= -\frac{1}{q}p_2\mathbf{b} - \frac{1}{r}p_3\mathbf{c}.$$

Similarly
$$\frac{\partial \mathbf{a}}{\partial v} = \frac{\partial}{\partial v}\left(\frac{\mathbf{r}_1}{p}\right) = \frac{1}{p}\mathbf{r}_{12} - \frac{1}{p^2}p_2\mathbf{r}_1$$

$$= \frac{1}{pq}q_1\mathbf{r}_2 \qquad \text{by (21)}$$

$$= \frac{1}{p}q_1\mathbf{b},$$

and
$$\frac{\partial \mathbf{a}}{\partial w} = \frac{1}{pr}r_1\mathbf{r}_3 = \frac{1}{p}r_1\mathbf{c},$$

with corresponding results for the derivatives of $\mathbf{b}$ and $\mathbf{c}$.

Ex. Prove the relations

$$\mathbf{r}_{11}{}^2 = p_1{}^2 + \frac{p^2}{q^2}p_2{}^2 + \frac{p^2}{r^2}p_3{}^2,$$

$$\mathbf{r}_{11} \bullet \mathbf{r}_{12} = p_1 p_2 - \frac{p}{q}p_2 q_1,$$

$$\mathbf{r}_{11} \bullet \mathbf{r}_{23} = -\frac{p}{q}p_2 q_3 - \frac{p}{r}p_3 r_2,$$

$$\mathbf{r}_{12} \bullet \mathbf{r}_{13} = p_2 p_3,$$

$$\mathbf{r}_{23}{}^2 = q_3{}^2 + r_2{}^2,$$

with similar results derivable from these by cyclic interchange of variables and suffixes.

111. Lamé's relations. The three parameters u, v, w are curvilinear coordinates of a point $\mathbf{r}$ in space. The length ds of an element of arc through the point is given by

$$ds^2 = d\mathbf{r}^2 = (\mathbf{r}_1 du + \mathbf{r}_2 dv + \mathbf{r}_3 dw)^2$$
$$= p^2 du^2 + q^2 dv^2 + r^2 dw^2,$$

since $\mathbf{r}_1$, $\mathbf{r}_2$, $\mathbf{r}_3$ are mutually perpendicular. The three functions p, q, r are not independent, but are connected by six differential equations, consisting of two groups of three. These were first deduced by Lamé*, and are called after him. We may write them

$$\left.\begin{array}{l} \dfrac{\partial}{\partial v}\left(\dfrac{r_2}{q}\right) + \dfrac{\partial}{\partial w}\left(\dfrac{q_3}{r}\right) + \dfrac{q_1 r_1}{p^2} = 0 \\[2ex] \dfrac{\partial}{\partial w}\left(\dfrac{p_3}{r}\right) + \dfrac{\partial}{\partial u}\left(\dfrac{r_1}{p}\right) + \dfrac{r_2 p_2}{q^2} = 0 \\[2ex] \dfrac{\partial}{\partial u}\left(\dfrac{q_1}{p}\right) + \dfrac{\partial}{\partial v}\left(\dfrac{p_2}{q}\right) + \dfrac{p_3 q_3}{r^2} = 0 \end{array}\right\} \quad \ldots\ldots\ldots\ldots(22),$$

* *Leçons sur les coordinées curvilignes et leurs diverses applications*, pp. 73–79 (1859).

and

$$p_{23} = \frac{p_2 q_3}{q} + \frac{p_3 r_2}{r}$$

$$q_{31} = \frac{q_3 r_1}{r} + \frac{q_1 p_3}{p} \left.\rule{0pt}{2em}\right\} \quad \dots\dots\dots\dots(23).$$

$$r_{12} = \frac{r_1 p_2}{p} + \frac{r_2 q_1}{q}$$

They may be proved by the method employed in establishing the Mainardi-Codazzi relations. Thus if in the identity

$$\frac{\partial}{\partial v} \mathbf{r}_{11} = \frac{\partial}{\partial u} \mathbf{r}_{12},$$

we substitute the values of $\mathbf{r}_{11}$ and $\mathbf{r}_{12}$ given by (20) and (21), and after differentiation substitute again the values of the second derivatives of $\mathbf{r}$ in terms of the first derivatives, we find an equation in which the coefficient of $\mathbf{r}_1$ vanishes identically, while the vanishing of the coefficients of $\mathbf{r}_2$ and $\mathbf{r}_3$ leads to the third equation of (22) and the first of (23). Similarly from the identity

$$\frac{\partial}{\partial w} \mathbf{r}_{22} = \frac{\partial}{\partial v} \mathbf{r}_{23},$$

we obtain the first of (22) and the second of (23); and from the identity

$$\frac{\partial}{\partial u} \mathbf{r}_{33} = \frac{\partial}{\partial w} \mathbf{r}_{31},$$

the second of (22) and the third of (23).

Moreover, just as the six fundamental magnitudes E, F, G; L, M, N, satisfying the Gauss characteristic equation and the Mainardi-Codazzi relations, determine a surface except as to position and orientation in space (Art. 44), so *the three functions p, q, r, satisfying Lamé's equations, determine a triply orthogonal system of surfaces except as to position and orientation in space.* But the proof of this theorem is beyond the scope of this book*.

Ex.† Given that the family $w = $ const. of a triply orthogonal system are surfaces of revolution, and that the curves $v = $ const. are meridians on these, examine the nature of the system.

On the surfaces $w = $ const., u and v are the parameters. Since the curves $v = $ const. are meridians they are also geodesics, and therefore E is a function of u only (Art. 47), the parametric curves being orthogonal. Thus $p_2 = 0$. From the first of (23) it then follows that either $r_2 = 0$ or $p_3 = 0$.

In the first case, since $p_2=0$ and $r_2=0$, (16) gives $\kappa_{vu}=0$ and $\kappa_{vw}=0$. Thus the surfaces $v=$const. are planes; and since they are meridian planes, the axes of the surfaces of revolution must coincide. The surfaces $w=$const. and $u=$const. are therefore those obtained by taking a family of plane curves and their orthogonal trajectories, and rotating their plane about a line in it as axis.

In the second case we have $p_3=0$, and therefore, in virtue of (17), $\kappa_{wu}=0$. Consequently the family of surfaces $w=$const. are developables, either circular cylinders or circular cones. Further, since $p_2=0$, $\kappa_{vu}=0$ by (16), and therefore the surfaces $v=$const. are also developables. And we have seen that

$$\kappa_u{}^2 = \kappa_{wu}{}^2 + \kappa_{vu}{}^2,$$

so that κ_u also vanishes. Thus the curves of parameter u are straight lines, and the surfaces $u=$const. parallel surfaces. These parallel surfaces are planes when the surfaces $w=$const. are cylinders.

*112. Theorems of Darboux. In conclusion we shall consider the questions whether any arbitrary family of surfaces forms part of a triply orthogonal system, and whether two orthogonal families of surfaces admit a third family orthogonal to both. As the answer to the second question supplies an answer to the first, we shall prove the following theorem due to Darboux:

A necessary and sufficient condition that two orthogonal families of surfaces admit a third family orthogonal to both is that their curves of intersection be lines of curvature on both.

Let the two orthogonal families of surfaces be

$$\left. \begin{array}{l} u\,(x,\,y,\,z) = \text{const.} \\ v\,(x,\,y,\,z) = \text{const.} \end{array} \right\} \dots\dots\dots\dots\dots(24).$$

Their normals are parallel to the vectors ∇u and ∇v. Denoting these gradients by **a** and **b** respectively, we have the condition of orthogonality of the surfaces,

$$\mathbf{a} \cdot \mathbf{b} = 0.$$

If there exists a third family of surfaces

$$w\,(x,\,y,\,z) = \text{const.} \dots\dots\dots\dots\dots\dots(25),$$

orthogonal to each of the above families, then any displacement $d\mathbf{r}$ tangential to (25) must be coplanar with **a** and **b**; that is

$$\mathbf{a} \times \mathbf{b} \cdot d\mathbf{r} = 0.$$

---

* This Art. is intended only for readers familiar with the formulae of advanced Vector Analysis. The differential invariants employed are three-parametric, and should not be confused with those of the following chapter.

The condition that this differential equation may admit an integral involving an arbitrary constant is

$$(\mathbf{a} \times \mathbf{b}) \cdot \nabla \times (\mathbf{a} \times \mathbf{b}) = 0,$$

which may be expanded

$$\mathbf{a} \times \mathbf{b} \cdot (\mathbf{b} \cdot \nabla \mathbf{a} - \mathbf{a} \cdot \nabla \mathbf{b} + \mathbf{a} \nabla \cdot \mathbf{b} - \mathbf{b} \nabla \cdot \mathbf{a}) = 0 \ \dots (26).$$

The scalar triple products from the last two terms vanish, owing to the repeated factor. Further, since $\mathbf{a} \cdot \mathbf{b} = 0$, it follows that

$$0 = \nabla (\mathbf{a} \cdot \mathbf{b}) = \mathbf{a} \cdot \nabla \mathbf{b} + \mathbf{b} \cdot \nabla \mathbf{a} + \mathbf{b} \times (\nabla \times \mathbf{a}) + \mathbf{a} \times (\nabla \times \mathbf{b}).$$

Again the last two terms vanish since

$$\nabla \times \mathbf{a} = \nabla \times \nabla u = 0$$

and

$$\nabla \times \mathbf{b} = \nabla \times \nabla v = 0.$$

Consequently

$$\mathbf{a} \cdot \nabla \mathbf{b} = - \mathbf{b} \cdot \nabla \mathbf{a}.$$

Substituting this value in (26), we have the condition

$$(\mathbf{a} \times \mathbf{b}) \cdot (\mathbf{a} \cdot \nabla \mathbf{b}) = 0 \ \dots \dots \dots \dots (27)$$

for the existence of a family of surfaces orthogonal to both the families (24).

Now consider a curve cutting the family of surfaces $u = $ const. orthogonally. A displacement $d\mathbf{r}$ along this curve is parallel to the vector $\mathbf{a}$ at the point and therefore, in virtue of the condition (27),

$$d\mathbf{r} \times \mathbf{b} \cdot (d\mathbf{r} \cdot \nabla \mathbf{b}) = 0,$$

which may be written

$$[d\mathbf{r}, \mathbf{b}, d\mathbf{b}] = 0.$$

Now the curve considered lies on a member of the family $v = $ const.; and, as $\mathbf{b}$ is normal to this surface, the last equation shows that the curve is a line of curvature. Thus the curves which cut the surfaces $u = $ const. orthogonally are lines of curvature on the surfaces $v = $ const. Hence their orthogonal trajectories on the latter are also lines of curvature. But these are the curves of intersection of the two families (24). Since these are lines of curvature on $v = $ const., and the two families cut at a constant angle, it follows from Joachimsthal's theorem that they are also lines of curvature on the surfaces $u = $ const., and Darboux's theorem is established.

We may now proceed to answer the other question, whether an arbitrary family of surfaces

$$u(x, y, z) = \text{const.} \ \dots \dots \dots \dots \dots (28)$$

forms part of a triply orthogonal system. If there is a second family of surfaces orthogonal to the above, their curves of intersection must be lines of curvature on $u = $ const. Hence, a family of lines of curvature on (28) must constitute a normal congruence if there are to be three orthogonal families.

Let **t** denote the *unit* tangent to a line of curvature on $u = $ const. Then the necessary and sufficient condition that the lines of curvature of this system should constitute a normal congruence is that **t** • dr admits an integral involving an arbitrary constant. The condition for this is

$$\mathbf{t} \cdot \nabla \times \mathbf{t} = 0 \quad \dots\dots\dots\dots\dots\dots(29).$$

As for the direction of **t** we observe that, if **n** is the *unit* normal to the surface $u = $ const., the tangent **t** to a line of curvature is parallel to the rate of change of **n** in that direction; that is to say, **t** is parallel to $d\mathbf{n}$, and therefore to $dr \cdot \nabla\mathbf{n}$. Hence, since dr has the direction of **t**,

$$\mathbf{t} \cdot \nabla\mathbf{n} = \lambda\mathbf{t},$$

where λ is a scalar factor. Thus **t** is expressible in terms of the first and second derivatives of u, and the equation (29) is therefore of the third order in these derivatives. Moreover the above analysis is reversible, and so we have Darboux's theorem:

In order that a family of surfaces $u(x, y, z) = $ const. may form part of a triply orthogonal system, it is necessary and sufficient that u should satisfy a certain partial differential equation of the third order.

Such a family of surfaces is called a *Lamé family*.

EXAMPLES XIV

1. Show that any family of spheres or planes, whose equation contains one parameter, can form part of a triply orthogonal system.

2. Show that a family of parallel surfaces is a Lamé family.

3. Prove the existence of a triply orthogonal system of spheres.

4. A necessary and sufficient condition that the surfaces $u = $ const. of a triply orthogonal system be parallel is that p be a function of u alone.

5. The curves $p = $ const. are *curves of equidistance* on a surface $u = $ const. between consecutive members of that family.

6. Examine the existence of a triply orthogonal system of minimal surfaces.

7. Prove that the equations (21), satisfied by $\mathbf{r}$, are also satisfied by $\mathbf{r}^2$.

8. Determine a triply orthogonal system of surfaces for which

$$p=1, \quad q=1, \quad r=Au+Bv+C,$$

where A, B, C are functions of w alone.

9. Prove that the surfaces

$$xy=uz^2, \quad x^2+y^2+z^2=v, \quad x^2+y^2+z^2=w\,(x^2-y^2)$$

are a triply orthogonal system.

10. Prove that the surfaces

$$yz=ax, \quad \sqrt{x^2+y^2}+\sqrt{x^2+z^2}=b, \quad \sqrt{x^2+y^2}-\sqrt{x^2+z^2}=c$$

cut one another orthogonally. Hence show that, on a hyperbolic paraboloid whose principal sections are equal parabolas, the sum or the difference of the distances of any point on a line of curvature from the two generators through the vertex is constant.

11. A triply orthogonal system of surfaces remains triply orthogonal after inversion (Art. 83).

12. Putting $p^2=a$, $q^2=b$, $r^2=c$, rewrite the equations (20) to (23) of the present chapter in terms of a, b, c and their derivatives*.

13. Calculate the first and second curvatures of the surfaces of a triply orthogonal system in terms of p, q, r; also in terms of a, b, c.

14. The *reciprocal* system of vectors to $\mathbf{r}_1$, $\mathbf{r}_2$, $\mathbf{r}_3$ of the present chapter is $\mathbf{l}$, $\mathbf{m}$, $\mathbf{n}$, where [*Elem. Vect. Anal.*, Art 47]

$$\mathbf{l}=\mathbf{r}_1/a, \quad \mathbf{m}=\mathbf{r}_2/b, \quad \mathbf{n}=\mathbf{r}_3/c.$$

Calculate the derivatives of these vectors in terms of $\mathbf{l}$, $\mathbf{m}$, $\mathbf{n}$, a, b, c.

* For orthogonal systems either of these notations is satisfactory; but, with triple systems generally, it is better to treat the squares and scalar products of the derivatives of $\mathbf{r}$ as the fundamental quantities. See Art. 128, or a recent paper by the author "On Triple Systems of Surfaces, and Non-Orthogonal Curvilinear Coordinates," *Proc. Royal Soc. Edin.* Vol. 46 (1926), pp. 194—205.

CHAPTER XII

DIFFERENTIAL INVARIANTS FOR A SURFACE

113. Point-functions. In this chapter we propose to give a brief account of the properties and uses of differential invariants for a surface. The "differential parameters" introduced by Beltrami and Darboux have long been employed in various parts of the subject. The author has shown, however, that these are only some of the scalar members of a family of both *vector* and *scalar* differential invariants *, which play an important part in geometry of surfaces, and in the discussion of physical problems connected with curved surfaces.

A quantity, which assumes one or more definite values at each point of a surface, is called a function of position or a *point-function* for the surface. If it has only one value at each point it is said to be uniform or single-valued. We shall be concerned with both scalar and vector point-functions; but in all cases the functions treated will be uniform. The value of the function at any point of the surface is determined by the coordinates u, v of that point; it is therefore a function of these variables.

114. Gradient of a scalar function. Consider first a scalar function of position, $\phi\,(u, v)$. We define the *gradient* or *slope* of the function at any point P as a vector quantity whose direction is that direction on the surface at P which gives the maximum arc-rate of increase of ϕ, and whose magnitude is this maximum rate of increase. There is no ambiguity about the direction; for it is the direction of increase, not decrease.

A curve $\phi =$ const. is called a *level curve* of the function. Let C, C' be two consecutive level curves, corresponding to the values ϕ and $\phi + d\phi$ of the function, where $d\phi$ is positive. Let PQ be an element of the orthogonal trajectory of the level curves, intercepted between C, C', and let dn be the length of this element. Let PR

---

* The theory of these invariants has been developed at some length by the author in a paper entitled: "On Differential Invariants in Geometry of Surfaces, with some applications to Mathematical Physics," *Quarterly Journal of Mathematics,* Vol. 50, pp. 230–269 (1925).

be an element of arc of another curve through P, cutting C' in R, and let ds be the length of PR. Then clearly PQ is the shortest distance from P to the curve C', and its direction is that which

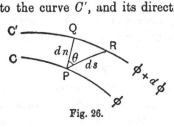

Fig. 26.

gives the maximum rate of increase of ϕ at P. Thus the gradient of ϕ at P has the direction PQ and the magnitude $d\phi/dn$. This vector will be denoted * by $\nabla\phi$ or grad ϕ. If $\mathbf{m}$ is the unit vector in the direction PQ, orthogonal to the curve $\phi = \text{const.}$, we have

$$\nabla\phi = \frac{d\phi}{dn}\,\mathbf{m} \quad\dots\dots\dots\dots\dots\dots(1).$$

And from the above definition it is clear that grad ϕ is independent of the choice of parameters u, v. It is itself a point-function for the surface.

The rate of increase of ϕ in the direction PR is given by

$$\frac{d\phi}{ds} = \frac{d\phi}{dn}\frac{dn}{ds} = \frac{d\phi}{dn}\cos\theta,$$

where θ is the inclination of PR to PQ. Thus *the rate of increase of ϕ in any direction along the surface is the resolved part of $\nabla\phi$ in that direction.* If $\mathbf{c}$ is the unit vector in the direction PR, the rate of increase of ϕ in this direction is therefore $\mathbf{c} \cdot \nabla\phi$. This may be called the *derivative* of ϕ in the direction of $\mathbf{c}$. If $d\mathbf{r}$ is the elementary vector PR we have $d\mathbf{r} = \mathbf{c}\,ds$; and therefore the change $d\phi$ in the function due to the displacement $d\mathbf{r}$ on the surface is given by

$$d\phi = \frac{d\phi}{ds}\,ds\,(\mathbf{c}\cdot\nabla\phi)$$

or $$d\phi = d\mathbf{r}\cdot\nabla\phi \quad\dots\dots\dots\dots\dots\dots(2).$$

From the definition of $\nabla\phi$ it is clear that the curves $\phi = \text{const.}$ will be parallels, provided the magnitude of $\nabla\phi$ is the same for all points on the same curve; that is to say, provided $(\nabla\phi)^2$ is a

---

* We shall borrow the notation and terminology of three-parametric differential invariants.

function of ϕ only. Hence *a necessary and sufficient condition that the curves $\phi = const.$ be parallels is that $(\nabla \phi)^2$ is a function of ϕ only.*

The curves $\psi = const.$ will be orthogonal trajectories of the curves $\phi = const.$ if the gradients of the two functions are everywhere perpendicular. Hence the *condition of orthogonality* of the two systems of curves is *

$$\nabla \phi \cdot \nabla \psi = 0.$$

Although the gradient of ϕ is independent of any choice of parameters, it will be convenient to have an expression for the function in terms of the selected coordinates u, v. This may be obtained as follows. If $\delta u, \delta v$ is an infinitesimal displacement along the curve $\phi(u, v) = const.$,

$$\phi_1 \delta u + \phi_2 \delta v = 0.$$

Hence a displacement du, dv orthogonal to this is given by (Art 24)

$$\frac{du}{dv} = \frac{G\phi_1 - F\phi_2}{E\phi_2 - F\phi_1}.$$

The vector

$$\mathbf{V} = (G\phi_1 - F\phi_2)\,\mathbf{r}_1 + (E\phi_2 - F\phi_1)\,\mathbf{r}_2$$

is therefore parallel to $\nabla \phi$. But the resolved part of this in the direction of $\mathbf{r}_1$ is equal to

$$\frac{1}{\sqrt{E}}\mathbf{r}_1 \cdot \mathbf{V} = (G\phi_1 - F\phi_2)\frac{E}{\sqrt{E}} + (E\phi_2 - F\phi_1)\frac{F}{\sqrt{E}}$$

$$= \frac{H^2}{\sqrt{E}}\phi_1 = H^2 \frac{1}{\sqrt{E}}\frac{\partial \phi}{\partial u},$$

---

* Beltrami's differential parameter of the first order, $\Delta_1 \phi$, is the square of the magnitude of $\nabla \phi$; that is

$$\Delta_1 \phi = (\nabla \phi)^2.$$

His mixed differential parameter of the first order, $\Delta_1 (\phi, \psi)$, is the scalar product of the gradients of ϕ and ψ; or

$$\Delta_1 (\phi, \psi) = \nabla \phi \cdot \nabla \psi.$$

Darboux's function $\Theta(\phi, \psi)$ is the magnitude of the vector product of $\nabla \phi$ and $\nabla \psi$; that is to say

$$\Theta(\phi, \psi)\,\mathbf{n} = \nabla \phi \times \nabla \psi.$$

The inclination θ of the curve $\psi = const.$ to the curve $\phi = const.$ is also the inclination of $\nabla \psi$ to $\nabla \phi$. And, since $\cos^2 \theta + \sin^2 \theta = 1$, it follows from the last two equations, on squaring and adding, that

$$\Delta_1{}^2 (\phi, \psi) + \Theta^2 (\phi, \psi) = (\nabla \phi)^2 (\nabla \psi)^2$$

$$= \Delta_1 \phi\, \Delta_1 \psi.$$

which is H^2 times the derivative of ϕ in the direction of $\mathbf{r}_1$. Hence grad ϕ is $\mathbf{V}/H^2$, or

$$\nabla\phi = \frac{(G\phi_1 - F\phi_2)}{H^2}\mathbf{r}_1 + \frac{(E\phi_2 - F\phi_1)}{H^2}\mathbf{r}_2 \ldots\ldots\ldots\ldots(3),$$

which is the required expression for the gradient.

We may regard this as the result obtained by operating on the function ϕ with the vectorial differential operator

$$\nabla = \frac{1}{H^2}\mathbf{r}_1\left(G\frac{\partial}{\partial u} - F\frac{\partial}{\partial v}\right) + \frac{1}{H^2}\mathbf{r}_2\left(E\frac{\partial}{\partial v} - F\frac{\partial}{\partial u}\right).$$

That this operator is invariant is clear from the definition of $\nabla\phi$ which is independent of parameters. The operator ∇ plays a fundamental part in the following argument; for all our invariants are expressible in terms of it. When *the parametric curves are orthogonal* it takes a simpler form. For then $F = 0$ and $H^2 = EG$, so that

$$\nabla = \frac{1}{E}\mathbf{r}_1\frac{\partial}{\partial u} + \frac{1}{G}\mathbf{r}_2\frac{\partial}{\partial v},$$

a form which will frequently be employed when it is desired to simplify the calculations.

Ex. Prove the following relations:

$$(\nabla\phi)^2 = \frac{1}{H^2}(E\phi_2^2 - 2F\phi_1\phi_2 + G\phi_1^2),$$

$$\nabla\phi \bullet \nabla\psi = \frac{1}{H^2}[E\phi_2\psi_2 - F(\phi_1\psi_2 + \phi_2\psi_1) + G\phi_1\psi_1],$$

$$\nabla\phi \times \nabla\psi = \frac{1}{H}(\phi_1\psi_2 - \phi_2\psi_1)\mathbf{n}.$$

115. Some applications. The gradients of the parameters u, v are given by

$$\nabla u = \frac{G}{H^2}\mathbf{r}_1 - \frac{F}{H^2}\mathbf{r}_2,$$

$$\nabla v = \frac{E}{H^2}\mathbf{r}_2 - \frac{F}{H^2}\mathbf{r}_1,$$

so that

$$\nabla u \times \nabla v = \frac{\mathbf{n}}{H},$$

and therefore

$$(\nabla u \times \nabla v)^2 = \frac{1}{H^2}.$$

Hence
$$(\nabla u)^2 = \frac{G}{H^2} = G\,(\nabla u \times \nabla v)^2,$$

$$\nabla u \cdot \nabla v = -\frac{F}{H^2} = -F\,(\nabla u \times \nabla v)^2,$$

$$(\nabla v)^2 = \frac{E}{H^2} = E\,(\nabla u \times \nabla v)^2.$$

If then it is desired to take as parametric variables any two functions ϕ, ψ, the corresponding values $\bar{E}, \bar{F}, \bar{G}, \bar{H}$ of the fundamental magnitudes are given by

$$\bar{H}^2 = \frac{1}{(\nabla\phi \times \nabla\psi)^2}$$

and
$$\bar{E} = \bar{H}^2(\nabla\psi)^2, \quad \bar{F} = -\bar{H}^2\nabla\phi \cdot \nabla\psi, \quad \bar{G} = \bar{H}^2(\nabla\phi)^2.$$

If the parametric curves are *orthogonal* we have simply

$$(\nabla u)^2 = \frac{1}{E}, \quad (\nabla v)^2 = \frac{1}{G}$$

In order that curves $u = $ const. may be a system of geodesic parallels, E must be a function of u only (Art. 56). Hence *a necessary and sufficient condition that the curves $\phi = $ const. be geodesic parallels is that $(\nabla\phi)^2$ be a function of ϕ only.* If the parameter ϕ is to measure the actual geodesic distance from a fixed parallel, we must have $(\nabla\phi)^2 = 1$.

The following application of the gradient will be required later. If C is a curve on the surface joining two points A, B, the definite integral from A to B of the resolved part of $\nabla\phi$ tangential to the curve is

$$\int_A^B \mathbf{t} \cdot \nabla\phi \, ds = \int_A^B d\mathbf{r} \cdot \nabla\phi = \int_A^B d\phi = \phi_B - \phi_A,$$

$\mathbf{t}$ being the unit tangent to the curve. Thus, if A is fixed, the definite integral is a point-function determined by the position of B. If ϕ is single-valued, and the definite integral is taken round a closed curve, ϕ_B becomes equal to ϕ_A and the integral vanishes. When the path of integration is closed we denote the fact by a small circle placed at the foot of the integral sign. Thus

$$\int_\circ \nabla\phi \cdot d\mathbf{r} = 0 \quad \dots\dots\dots\dots\dots\dots (4).$$

Conversely, suppose that the vector **F** is tangential to the surface. and that $\int_0 \mathbf{F} \cdot d\mathbf{r}$ vanishes for *every* closed curve drawn on the surface. Then $\int_A^B \mathbf{F} \cdot d\mathbf{r}$ must be the same for all paths joining A and B. If then A is fixed, the value of the integral is a point-function ϕ determined by the position of B. Hence, for any small displacement $d\mathbf{r}$ of B, we have

$$\mathbf{F} \cdot d\mathbf{r} = d\phi = \nabla\phi \cdot d\mathbf{r}.$$

This is true for all values of $d\mathbf{r}$ tangential to the surface. Hence, since **F** and $\nabla\phi$ are both tangential to the surface they must be equal, and we have the theorem:

If a vector point-function **F** *is everywhere tangential to the surface, and* $\int_0 \mathbf{F} \cdot d\mathbf{r}$ *vanishes for every closed curve drawn on the surface, then* **F** *is the gradient of some scalar point-function.*

116. Divergence of a vector. The operator ∇ may be applied to a vector function **F** in different ways. One of these leads to a *scalar* differential invariant, which we shall call the *divergence* of **F** and shall denote by div **F** or $\nabla \cdot \mathbf{F}$. We define it by the equation

$$\text{div } \mathbf{F} = \nabla \cdot \mathbf{F}$$
$$= \frac{1}{H^2}\mathbf{r}_1 \cdot \left(G\frac{\partial \mathbf{F}}{\partial u} - F\frac{\partial \mathbf{F}}{\partial v}\right) + \frac{1}{H^2}\mathbf{r}_2 \cdot \left(E\frac{\partial \mathbf{F}}{\partial v} - F\frac{\partial \mathbf{F}}{\partial u}\right).$$

That this is invariant with respect to the choice of parameters may be shown by actual transformation from one pair to another. But it is unnecessary to do this, as the invariant property will follow directly from another expression that will shortly be found for div **F**, which is entirely independent of coordinates.

To illustrate the importance of the divergence function, consider the divergence of the unit normal **n** to the surface. Thus

$$\text{div } \mathbf{n} = \frac{1}{H^2}\mathbf{r}_1 \cdot (G\mathbf{n}_1 - F\mathbf{n}_2) + \frac{1}{H^2}\mathbf{r}_2 \cdot (E\mathbf{n}_2 - F\mathbf{n}_1)$$

and, on substituting the values of $\mathbf{n}_1$ and $\mathbf{n}_2$ given in Art. 27, we find

$$\text{div } \mathbf{n} = -\frac{1}{H^2}(EN - 2FM + GL)$$
$$= -J \dotfill (5),$$

where J as usual denotes the first curvature. Hence:

The first curvature of a surface is the negative of the divergence of the unit normal *.

If R is a scalar point-function we may find the divergence of $R\mathbf{n}$ in the same manner. Then, remembering that $\mathbf{n}$ is perpendicular to $\mathbf{r}_1$ and $\mathbf{r}_2$ we find

$$\operatorname{div} R\mathbf{n} = R \operatorname{div} \mathbf{n} = -JR \quad \dots\dots\dots\dots(6).$$

The first curvature of a minimal surface vanishes identically. Hence *minimal surfaces* are characterised by the relations

$$\operatorname{div} \mathbf{n} = 0$$

or $$\operatorname{div} R\mathbf{n} = 0.$$

It will be convenient to have an expression for the divergence of a vector in terms of the components of the vector. Suppose, for instance, that

$$\mathbf{F} = P\mathbf{r}_1 + Q\mathbf{r}_2 + R\mathbf{n}.$$

Then clearly, from the definition,

$$\operatorname{div} \mathbf{F} = \operatorname{div}(P\mathbf{r}_1) + \operatorname{div}(Q\mathbf{r}_2) + \operatorname{div}(R\mathbf{n}).$$

The value of the last term has already been found. Consider the first term. We have

$$\operatorname{div} P\mathbf{r}_1 = \frac{1}{H^2}\mathbf{r}_1 \cdot [G(P_1\mathbf{r}_1 + P\mathbf{r}_{11}) - F(P_2\mathbf{r}_1 + P\mathbf{r}_{12})]$$

$$+ \frac{1}{H^2}\mathbf{r}_2 \cdot [E(P_2\mathbf{r}_1 + P\mathbf{r}_{12}) - F(P_1\mathbf{r}_1 + P\mathbf{r}_{11})]$$

$$= P_1 + \frac{P}{2H^2}\left[E\frac{\partial}{\partial u}(\mathbf{r}_2{}^2) + G\frac{\partial}{\partial u}(\mathbf{r}_1{}^2) - 2F\frac{\partial}{\partial u}(\mathbf{r}_1 \cdot \mathbf{r}_2)\right].$$

* The reader who is familiar with the three-parametric divergence will recognise that (5) is true also for this value of **div n**. For the two functions differ only by the term $\mathbf{n} \cdot \dfrac{\partial \mathbf{n}}{\partial n}$, where $\dfrac{\partial}{\partial n}$ denotes differentiation along the normal. But **n** is perpendicular to its derivative, because it is a vector of constant length. Thus the value of the extra term is zero.

This provides a formula for the first curvature of a surface of the family

$$F(x, y, z) = \text{const.}$$

For the unit normal is the vector

$$\mathbf{n} = \left(\frac{1}{\mu}\frac{\partial F}{\partial x}, \ \frac{1}{\mu}\frac{\partial F}{\partial y}, \ \frac{1}{\mu}\frac{\partial F}{\partial z}\right),$$

where $$\mu^2 = \left(\frac{\partial F}{\partial x}\right)^2 + \left(\frac{\partial F}{\partial y}\right)^2 + \left(\frac{\partial F}{\partial z}\right)^2.$$

Hence except for sign, which is arbitrary without some convention,

$$J = \operatorname{div} \mathbf{n}$$

$$= \frac{\partial}{\partial x}\left(\frac{1}{\mu}\frac{\partial F}{\partial x}\right) + \frac{\partial}{\partial y}\left(\frac{1}{\mu}\frac{\partial F}{\partial y}\right) + \frac{\partial}{\partial z}\left(\frac{1}{\mu}\frac{\partial F}{\partial z}\right).$$

Thus $\qquad \operatorname{div} P\mathbf{r}_1 = P_1 + \dfrac{P}{2H^2} \dfrac{\partial}{\partial u} (EG - F^2)$

$$= \dfrac{1}{H} \dfrac{\partial}{\partial u} (HP).$$

Similarly we find $\qquad \operatorname{div} Q\mathbf{r}_2 = \dfrac{1}{H} \dfrac{\partial}{\partial v} (HQ).$

Hence the complete formula

$$\operatorname{div} \mathbf{F} = \dfrac{1}{H} \left[\dfrac{\partial}{\partial u} (HP) + \dfrac{\partial}{\partial v} (HQ) \right] - JR \dots\dots\dots\dots(7).$$

An important particular case is that in which the vector $\mathbf{F}$ is equal to the gradient of a scalar function ϕ. The divergence of the gradient* of ϕ is $\nabla \cdot \nabla \phi$ and will be written $\nabla^2 \phi$. The operator ∇^2 is analogous to the *Laplacian* of three parameters. Inserting in (7) the values of P and Q found from (3), we have

$$\nabla^2 \phi = \dfrac{1}{H} \dfrac{\partial}{\partial u} \left(\dfrac{G\phi_1 - F\phi_2}{H} \right) + \dfrac{1}{H} \dfrac{\partial}{\partial v} \left(\dfrac{E\phi_2 - F\phi_1}{H} \right) \dots\dots(8).$$

When the parametric curves are orthogonal this takes the simpler form

$$\nabla^2 \phi = \dfrac{1}{\sqrt{EG}} \left[\dfrac{\partial}{\partial u} \left(\phi_1 \sqrt{\dfrac{G}{E}} \right) + \dfrac{\partial}{\partial v} \left(\phi_2 \sqrt{\dfrac{E}{G}} \right) \right] \dots\dots(8').$$

117. Isometric parameters. From the last result it follows that, when the parametric curves are orthogonal,

$$\nabla^2 u = \dfrac{1}{2G} \dfrac{\partial}{\partial u} \left(\dfrac{G}{E} \right)$$

$$= \tfrac{1}{2} (\nabla v)^2 \dfrac{\partial}{\partial u} \left(\dfrac{G}{E} \right)$$

by Art. 115. If then u, v are isometric parameters, so that

$$ds^2 = \lambda (du^2 + dv^2),$$

the quotient G/E is constant, and therefore $\nabla^2 u = 0$. Conversely if $F = 0$ and also $\nabla^2 u = 0$ it follows from the above equation that G/E is constant, or a function of v only, so that

$$ds^2 = \lambda (du^2 + V dv^2).$$

We may then take $\int \sqrt{V} \, dv$ for a new second parameter, and our parameters are thus isometric. Hence the theorem:

---

* The invariant $\nabla^2 \phi$ is identical with Beltrami's differential parameter of the second order, $\Delta_2 \phi$.

A necessary and sufficient condition that u be the isometric parameter of one family of an isometric system is that $\nabla^2 u = 0$.

Again, if the square of the linear element is of the form

$$ds^2 = \lambda \, (U \, du^2 + V \, dv^2),$$

where U is a function of u only, and V a function of v only, the parametric curves are isometric lines (Art. 39). Also

$$\nabla^2 u = \frac{1}{2G} \frac{\partial}{\partial u} \left(\frac{V}{U} \right) = - \frac{V}{2G} \frac{U_1}{U^2} = - \frac{1}{2E} \frac{U_1}{U}$$

$$= - \tfrac{1}{2} (\nabla u)^2 f(u),$$

where $f(u)$ is a function of u only. Conversely if $\nabla^2 u / (\nabla u)^2$ is a function of u only, the curves $u = $ const. and their orthogonal trajectories $v = $ const. form an isometric system. For since

$$\nabla^2 u = \tfrac{1}{2} (\nabla v)^2 \frac{\partial}{\partial u} \left(\frac{G}{E} \right)$$

it follows, in virtue of the last equation, that

$$\frac{\partial}{\partial u} \left(\frac{G}{E} \right) = \frac{2 \nabla^2 u}{(\nabla v)^2} = - \frac{(\nabla u)^2}{(\nabla v)^2} f(u)$$

$$= - \frac{G}{E} f(u).$$

Thus

$$\frac{\partial}{\partial u} \log \frac{G}{E} = - f(u),$$

and therefore

$$\frac{\partial^2}{\partial u \, \partial v} \log \frac{G}{E} = 0,$$

showing that the parametric curves are isometric. Thus:

A necessary and sufficient condition that a family of curves $u = $ const. and their orthogonal trajectories form an isometric system is that $\nabla^2 u / (\nabla u)^2$ be a function of u.

118. Curl of a vector. The operator ∇ may be applied to a vector function $\mathbf{F}$ in such a way as to give a *vector* differential invariant. We shall call this the *curl* or *rotation* of $\mathbf{F}$ and denote it by curl $\mathbf{F}$ or $\nabla \times \mathbf{F}$. It is defined by the equation

$$\text{curl } \mathbf{F} = \nabla \times \mathbf{F}$$

$$= \frac{1}{H^2} \mathbf{r}_1 \times \left(G \frac{\partial \mathbf{F}}{\partial u} - F \frac{\partial \mathbf{F}}{\partial v} \right) + \frac{1}{H^2} \mathbf{r}_2 \times \left(E \frac{\partial \mathbf{F}}{\partial v} - F \frac{\partial \mathbf{F}}{\partial u} \right).$$

The invariant property of this function will appear from another expression that will be found for it, entirely independent of co-ordinates.

Consider first the curl of a vector $R\mathbf{n}$ normal to the surface. Then

$$\operatorname{curl} R\mathbf{n} = \frac{1}{H^2}\mathbf{r}_1 \times [G(R_1\mathbf{n} + R\mathbf{n}_1) - F(R_2\mathbf{n} + R\mathbf{n}_2)]$$
$$+ \frac{1}{H^2}\mathbf{r}_2 \times [E(R_2\mathbf{n} + R\mathbf{n}_2) - F(R_1\mathbf{n} + R\mathbf{n}_1)].$$

On substituting the values of $\mathbf{n}_1$ and $\mathbf{n}_2$ as given in Art. 27 we find that the terms involving these derivatives disappear; and the formula may then be written

$$\operatorname{curl} R\mathbf{n} = \nabla R \times \mathbf{n} \quad \dots\dots\dots\dots\dots\dots(9).$$

We may notice that this vector is perpendicular to $\mathbf{n}$ and there-fore tangential to the surface. If R is constant, ∇R vanishes, so that *the curl of any normal vector of constant length vanishes identi-cally.* In particular the curl of the unit normal is zero: or

$$\nabla \times \mathbf{n} = \operatorname{curl} \mathbf{n} = 0 \quad \dots\dots\dots\dots\dots\dots(10).$$

And we may here notice also that the curl of the position vector $\mathbf{r}$ of the current point on the surface vanishes identically: or

$$\nabla \times \mathbf{r} = \operatorname{curl} \mathbf{r} = 0 \quad \dots\dots\dots\dots\dots\dots(11).$$

As in the case of the divergence we may find an expansion for curl $\mathbf{F}$ in terms of the components of $\mathbf{F}$. For, if

$$\mathbf{F} = P\mathbf{r}_1 + Q\mathbf{r}_2 + R\mathbf{n},$$

we have $\quad \operatorname{curl} \mathbf{F} = \operatorname{curl} P\mathbf{r}_1 + \operatorname{curl} Q\mathbf{r}_2 + \operatorname{curl} R\mathbf{n}.$

The value of the last term has already been found. The first term is equal to

$$\operatorname{curl} P\mathbf{r}_1 = \frac{1}{H^2}\mathbf{r}_1 \times [G(P_1\mathbf{r}_1 + P\mathbf{r}_{11}) - F(P_2\mathbf{r}_1 + P\mathbf{r}_{12})]$$
$$+ \frac{1}{H^2}\mathbf{r}_2 \times [E(P_2\mathbf{r}_1 + P\mathbf{r}_{12}) - F(P_1\mathbf{r}_1 + P\mathbf{r}_{11})].$$

On substituting the values of $\mathbf{r}_{11}$ and $\mathbf{r}_{12}$ as given by Gauss's formulae (Art. 41) we find on reduction

$$\operatorname{curl} P\mathbf{r}_1 = \frac{1}{H}\left[\frac{\partial}{\partial u}(FP) - \frac{\partial}{\partial v}(EP)\right]\mathbf{n} + \frac{P}{H}(M\mathbf{r}_1 - L\mathbf{r}_2).$$

Similarly

$$\operatorname{curl} Q\mathbf{r}_2 = \frac{1}{H}\left[\frac{\partial}{\partial u}(GQ) - \frac{\partial}{\partial v}(FQ)\right]\mathbf{n} + \frac{Q}{H}(N\mathbf{r}_1 - M\mathbf{r}_2).$$

Taking the sum of the three results, we have the complete formula

$$\operatorname{curl} \mathbf{F} = \frac{1}{H}\left[\frac{\partial}{\partial u}(FP+GQ) - \frac{\partial}{\partial v}(EP+FQ)\right]\mathbf{n} + \frac{1}{H}(PM+QN)\mathbf{r}_1$$

$$-\frac{1}{H}(PL+QM)\mathbf{r}_2 + \nabla R \times \mathbf{n} \ldots\ldots\ldots\ldots(12).$$

One important consequence of this may be noticed. If we put $\mathbf{F} = \nabla\phi$, and substitute the corresponding values of P and Q as found from (3), the coefficient of $\mathbf{n}$ vanishes identically. Thus:

The curl of the gradient of a scalar point-function is tangential to the surface.

An equally important converse will soon be proved, viz.: If both $\mathbf{F}$ and curl $\mathbf{F}$ are tangential to the surface, then $\mathbf{F}$ is the gradient of some scalar function.

119. Vector functions (*cont.*). We have seen that $\mathbf{c} \cdot \nabla\phi$ is the derivative of ϕ in the direction of the unit vector $\mathbf{c}$. The same operator $\mathbf{c} \cdot \nabla$ may be applied to a vector function, giving the *derivative* of the vector in that direction. Thus

$$\mathbf{c} \cdot \nabla\mathbf{F} = \frac{1}{H^2}\mathbf{c} \cdot \mathbf{r}_1\left(G\frac{\partial\mathbf{F}}{\partial u} - F\frac{\partial\mathbf{F}}{\partial v}\right) + \frac{1}{H^2}\mathbf{c} \cdot \mathbf{r}_2\left(E\frac{\partial\mathbf{F}}{\partial v} - F\frac{\partial\mathbf{F}}{\partial u}\right)$$

is the derivative of $\mathbf{F}$ in the direction of $\mathbf{c}$. As a particular case put $\mathbf{c} = \mathbf{r}_1/\sqrt{E}$ and we find for the derivative of $\mathbf{F}$ in this direction

$$\frac{1}{H^2\sqrt{E}}[E(G\mathbf{F}_1 - F\mathbf{F}_2) + F(E\mathbf{F}_2 - F\mathbf{F}_1)]$$

$$= \frac{1}{\sqrt{E}}\mathbf{F}_1 = \frac{1}{\sqrt{E}}\frac{\partial\mathbf{F}}{\partial u}$$

as required. Though this interpretation of $\mathbf{c} \cdot \nabla\mathbf{F}$ as the "rate of change" of $\mathbf{F}$ in the direction of $\mathbf{c}$ is applicable only when $\mathbf{c}$ is tangential to the surface, we define the function $\mathbf{c} \cdot \nabla\mathbf{F}$ for all values of $\mathbf{c}$ by the above equation.

Similarly the operator ∇^2 defined by (8) may be applied to a vector point-function, giving a vector differential invariant of the second order. As illustrations we shall calculate the values of $\nabla^2\mathbf{r}$ and $\nabla^2\mathbf{n}$, where $\mathbf{r}$ and $\mathbf{n}$ have their usual meanings. We lose no generality by taking orthogonal parametric curves. Then, in virtue of (8'),

$$\nabla^2 \mathbf{r} = \frac{1}{H} \frac{\partial}{\partial u} \left(\mathbf{r}_1 \sqrt{\frac{G}{E}} \right) + \frac{1}{H} \frac{\partial}{\partial v} \left(\mathbf{r}_2 \sqrt{\frac{E}{G}} \right)$$

$$= \frac{1}{\sqrt{EG}} \left[\frac{\partial}{\partial u} \left(\sqrt{\frac{G}{E}} \right) \mathbf{r}_1 + \sqrt{\frac{G}{E}} \left(L\mathbf{n} + \frac{E_1}{2E} \mathbf{r}_1 - \frac{E_2}{2G} \mathbf{r}_2 \right) \right.$$

$$\left. + \frac{\partial}{\partial v} \left(\sqrt{\frac{E}{G}} \right) \mathbf{r}_2 + \sqrt{\frac{E}{G}} \left(N\mathbf{n} - \frac{G_1}{2E} \mathbf{r}_1 + \frac{G_2}{2G} \mathbf{r}_2 \right) \right]$$

$$= \left(\frac{L}{E} + \frac{N}{G} \right) \mathbf{n}.$$

Thus $\qquad\qquad\qquad \nabla^2 \mathbf{r} = J\mathbf{n}$ ....(13)
is the required relation.

Cor. 1. From this equation and (6) we have

$$\nabla \cdot \nabla^2 \mathbf{r} = - J^2.$$

Cor. 2. The current point $\mathbf{r}$ on a minimal surface satisfies the equation

$$\nabla^2 \mathbf{r} = 0.$$

Next consider the function $\nabla^2 \mathbf{n}$. For simplicity in calculation we shall take the lines of curvature as parametric curves. Then $F = 0$, $M = 0$ and

$$\mathbf{n}_1 = - \frac{L}{E} \mathbf{r}_1, \quad \mathbf{n}_2 = - \frac{N}{G} \mathbf{r}_2.$$

Hence by (8′)

$$\nabla^2 \mathbf{n} = - \frac{1}{\sqrt{EG}} \left[\frac{\partial}{\partial u} \left(\frac{L}{E} \sqrt{\frac{G}{E}} \mathbf{r}_1 \right) + \frac{\partial}{\partial v} \left(\frac{N}{G} \sqrt{\frac{E}{G}} \mathbf{r}_2 \right) \right].$$

Then in virtue of Gauss's formulae for $\mathbf{r}_{11}$ and $\mathbf{r}_{22}$, and the Mainardi-Codazzi relations, this reduces to

$$\nabla^2 \mathbf{n} = - \left(\frac{L^2}{E^2} + \frac{N^2}{G^2} \right) \mathbf{n} - \left(\frac{1}{E} \mathbf{r}_1 \frac{\partial}{\partial u} + \frac{1}{G} \mathbf{r}_2 \frac{\partial}{\partial v} \right) \left(\frac{L}{E} + \frac{N}{G} \right)$$

$$= - [J^2 - 2K] \mathbf{n} - \nabla J \qquad ....(14),$$

which is the required formula, K as usual denoting the second curvature.

Since ∇J is tangential to the surface, by forming the scalar product of each member of (14) with $\mathbf{n}$ we deduce

$$\left. \begin{aligned} 2K &= \mathbf{n} \cdot \nabla^2 \mathbf{n} + J^2 \\ &= \mathbf{n} \cdot \nabla^2 \mathbf{n} + (\nabla \cdot \mathbf{n})^2 \end{aligned} \right\} \quad ....(15),$$

which may also be written

$$2K = \mathbf{n} \cdot \nabla^2 \mathbf{n} - \nabla \cdot \nabla^2 \mathbf{r} ....(16).$$

Thus *the second curvature is a differential invariant of* **n** *of the second order.*

120. Formulae of expansion. If ϕ denotes a scalar point-function, and **U**, **V** vector point-functions, we require expressions for the divergence and the curl of the functions $\phi\mathbf{U}$ and $\mathbf{U} \times \mathbf{V}$, in terms of the differential invariants of the separate functions. The formulae required may be expressed*

$$\nabla \cdot (\phi\mathbf{U}) = \nabla\phi \cdot \mathbf{U} + \phi\nabla \cdot \mathbf{U} \quad\ldots\ldots\ldots\ldots(17),$$

$$\nabla \times (\phi\mathbf{U}) = \nabla\phi \times \mathbf{U} + \phi\nabla \times \mathbf{U} \quad\ldots\ldots\ldots\ldots(18),$$

$$\nabla \cdot (\mathbf{U} \times \mathbf{V}) = \mathbf{V} \cdot \nabla \times \mathbf{U} - \mathbf{U} \cdot \nabla \times \mathbf{V} \quad\ldots\ldots\ldots(19).$$

For brevity of expression in the proof of these we may suppose the parametric curves orthogonal. Then

$$\nabla \cdot (\phi\mathbf{U}) = \frac{1}{E}\mathbf{r}_1 \cdot (\phi_1\mathbf{U} + \phi\mathbf{U}_1) + \frac{1}{G}\mathbf{r}_2 \cdot (\phi_2\mathbf{U} + \phi\mathbf{U}_2)$$

$$= \left(\frac{1}{E}\mathbf{r}_1\phi_1 + \frac{1}{G}\mathbf{r}_2\phi_2\right) \cdot \mathbf{U} + \phi\left(\frac{1}{E}\mathbf{r}_1 \cdot \mathbf{U}_1 + \frac{1}{G}\mathbf{r}_2 \cdot \mathbf{U}_2\right)$$

$$= \nabla\phi \cdot \mathbf{U} + \phi\nabla \cdot \mathbf{U},$$

which proves (17); and (18) may be established in a similar manner. In the case of (19) we have

$$\nabla \cdot (\mathbf{U} \times \mathbf{V}) = \frac{1}{E}\mathbf{r}_1 \cdot (\mathbf{U}_1 \times \mathbf{V} + \mathbf{U} \times \mathbf{V}_1) + \frac{1}{G}\mathbf{r}_2 \cdot (\mathbf{U}_2 \times \mathbf{V} + \mathbf{U} \times \mathbf{V}_2).$$

Then since the dot and the cross in a scalar triple product may be interchanged, provided the cyclic order of the factors is maintained, we have

$$\nabla \cdot (\mathbf{U} \times \mathbf{V}) = \left(\frac{1}{E}\mathbf{r}_1 \times \mathbf{U}_1 + \frac{1}{G}\mathbf{r}_2 \times \mathbf{U}_2\right) \cdot \mathbf{V} - \mathbf{U} \cdot \left(\frac{1}{E}\mathbf{r}_1 \times \mathbf{V}_1 + \ldots\right)$$

$$= \mathbf{V} \cdot \nabla \times \mathbf{U} - \mathbf{U} \cdot \nabla \times \mathbf{V}$$

as required.

As examples, apply (17) and (18) to the function $R\mathbf{n}$. Then

$$\operatorname{div} R\mathbf{n} = \nabla R \cdot \mathbf{n} + R\operatorname{div}\mathbf{n}.$$

But ∇R is perpendicular to **n**, and $\operatorname{div}\mathbf{n} = -J$. Hence

$$\operatorname{div} R\mathbf{n} = -JR$$

---

* For other formulae see § 7 of the author's paper "On Differential Invariants etc.," already referred to, or Examples XV below.

as previously found. Similarly

$$\operatorname{curl} R\mathbf{n} = \nabla R \times \mathbf{n} + R \operatorname{curl} \mathbf{n}.$$

But curl $\mathbf{n} = 0$, and therefore

$$\operatorname{curl} R\mathbf{n} = \nabla R \times \mathbf{n},$$

agreeing with (9). Again

$$\operatorname{div} \operatorname{curl} R\mathbf{n} = \operatorname{div} (\nabla R \times \mathbf{n})$$
$$= \mathbf{n} \cdot \nabla \times \nabla R - \nabla R \cdot \nabla \times \mathbf{n}.$$

Now each of these terms vanishes; the first because curl grad R is tangential to the surface (Art. 118), and the second because curl $\mathbf{n} = 0$. Hence

$$\operatorname{div} \operatorname{curl} R\mathbf{n} = 0 \quad\dots\dots\dots\dots\dots(20).$$

Thus *if a vector function is everywhere normal to the surface, the divergence of its curl vanishes identically.*

Ex. 1. Show that $\operatorname{div} \mathbf{r} = 2.$

Ex. 2. Show that $\operatorname{div} J\mathbf{r} = \mathbf{r} \cdot \nabla J + 2J$

and $\operatorname{curl} J\mathbf{r} = \nabla J \times \mathbf{r}.$

Ex. 3. Prove the formulae:

$$\nabla (\phi\psi) = \phi\nabla\psi + \psi\nabla\phi,$$
$$\nabla^2 (\phi\psi) = \phi\nabla^2\psi + 2\nabla\phi \cdot \nabla\psi + \psi\nabla^2\phi,$$
$$\nabla \times \nabla (\phi\psi) = \phi\nabla \times \nabla\psi + \psi\nabla \times \nabla\phi.$$

121. Geodesic curvature. Take any orthogonal system of parametric curves, and let $\mathbf{a}, \mathbf{b}$ be unit vectors in the directions of $\mathbf{r}_1$ and $\mathbf{r}_2$, so that

$$\mathbf{a} = \frac{\mathbf{r}_1}{\sqrt{E}}, \quad \mathbf{b} = \frac{\mathbf{r}_2}{\sqrt{G}}.$$

Then $\mathbf{a}, \mathbf{b}, \mathbf{n}$ form a right-handed system of mutually perpendicular unit vectors, such that

$$\mathbf{a} \times \mathbf{b} = \mathbf{n}, \quad \mathbf{b} \times \mathbf{n} = \mathbf{a}, \quad \mathbf{n} \times \mathbf{a} = \mathbf{b}.$$

The vector curvature of the parametric curve $v = \text{const.}$ is the arc-rate of change of $\mathbf{a}$ in this direction, which is equal to $\frac{1}{\sqrt{E}}\frac{\partial \mathbf{a}}{\partial u}$. But

$$\frac{\partial \mathbf{a}}{\partial u} = \frac{\partial}{\partial u}\left(\frac{\mathbf{r}_1}{\sqrt{E}}\right) = \frac{1}{\sqrt{E}}\mathbf{r}_{11} - \frac{E_1}{2E\sqrt{E}}\mathbf{r}_1$$
$$= \frac{L}{\sqrt{E}}\mathbf{n} - \frac{E_2}{2\sqrt{EG}}\mathbf{b}$$

by Art. 41. Hence the vector curvature of the curve $v =$ const. is

$$\frac{L}{E}\mathbf{n} - \frac{E_2}{2E\sqrt{G}}\,\mathbf{b}.$$

The normal component of this, L/E, is the curvature of the geodesic tangent, or the normal curvature of the surface in the direction of **a**. The tangential component is the curvature relative to the geodesic, or the *geodesic curvature* (Art. 53). Since this is regarded as positive when the relative curvature is in the positive sense for a rotation about the normal, the geodesic curvature of $v =$ const. is the coefficient of **b** in the above expression; or

$$\kappa_g = -\frac{E_2}{2E\sqrt{G}}.$$

Now the divergence of **b** is, by (7),

$$\operatorname{div}\left(\frac{\mathbf{r}_2}{\sqrt{G}}\right) = \frac{1}{H}\frac{\partial}{\partial v}\left(\frac{H}{\sqrt{G}}\right) = \frac{1}{\sqrt{EG}}\frac{\partial}{\partial v}(\sqrt{E}) = \frac{E_2}{2E\sqrt{G}}$$

$$= -\kappa_g.$$

Hence the geodesic curvature of the parametric curve $v =$ const. is the negative of the divergence of the unit vector **b**. But the parametric curves $v =$ const. may be chosen arbitrarily. Hence the theorem:

Given a family of curves on the surface, with an assigned positive direction along the curves, the geodesic curvature of a member of the family is the negative of the divergence of the unit vector tangential to the surface and orthogonal to the curve, whose direction is obtained by a positive rotation of one right angle (about the normal) from the direction of the curve.

The same result could have been obtained by considering the curve $u =$ const. In this case the vector curvature is

$$\frac{1}{\sqrt{G}}\frac{\partial \mathbf{b}}{\partial v} = \frac{N}{G}\mathbf{n} - \frac{G_1}{2G\sqrt{E}}\mathbf{a}.$$

The geodesic curvature of $u =$ const. is the resolved part of this in the direction of $-\mathbf{a}$, which is the direction obtained by a positive rotation of one right angle from **b** about the normal. Hence

$$\kappa_g = \frac{G_1}{2G\sqrt{E}}.$$

But this is equal to div **a**, and is therefore the negative of div $(-\mathbf{a})$, as required by the above theorem.

Bonnet's formula for the geodesic curvature of the curve $\phi(u, v) = $ const. (Art. 55) follows immediately from this theorem. For the unit vector orthogonal to the curve is $\nabla\phi / |\nabla\phi|$. But

$$|\nabla\phi| = \frac{1}{H} \sqrt{(E\phi_2^2 - 2F\phi_1\phi_2 + G\phi_1^2)},$$

and therefore

$$\kappa_g = \mathrm{div} \frac{\nabla\phi}{|\nabla\phi|} \quad \dots\dots\dots\dots\dots \dots\dots\dots\dots(21)$$

$$= \frac{1}{H}\left[\frac{\partial}{\partial u}\left(\frac{G\phi_1 - F\phi_2}{\sqrt{(E\phi_2^2 - 2F\phi_1\phi_2 + G\phi_1^2)}} \right) \right.$$
$$\left. + \frac{\partial}{\partial v}\left(\frac{E\phi_2 - F\phi_1}{\sqrt{(E\phi_2^2 - 2F\phi_1\phi_2 + G\phi_1^2)}} \right) \right].$$

The sign is indeterminate unless one direction along the curve is taken as the positive direction.

Another formula for the geodesic curvature of a curve may be deduced from the above theorem. For if **t** is the unit tangent to the curve, the unit vector orthogonal to the curve in the sense indicated is **n** × **t**. Hence the required geodesic curvature is

$$\kappa_g = -\,\mathrm{div}\,(\mathbf{n} \times \mathbf{t})$$
$$= \mathbf{n} \cdot \nabla \times \mathbf{t} - \mathbf{t} \cdot \nabla \times \mathbf{n}$$

by (19). But curl **n** is zero, and the last term vanishes, giving

$$\kappa_g = \mathbf{n} \cdot \mathrm{curl}\,\mathbf{t} \quad \dots\dots\dots\dots\dots\dots(22).$$

Hence the theorem:

Given a family of curves on the surface, with an assigned positive direction along the curves, the geodesic curvature of a member of the family is the normal resolute of the curl of the unit tangent.

We may observe in passing that, since the parametric curves are orthogonal, the curl of the unit tangent **a** to the line $v = $ const. is, by (12),

$$\mathrm{curl}\,\mathbf{a} = \frac{M}{\sqrt{EG}}\,\mathbf{a} - \frac{L}{E}\,\mathbf{b} - \frac{E_2}{2E\sqrt{G}}\,\mathbf{n},$$

and similarly $\quad \mathrm{curl}\,\mathbf{b} = \frac{N}{G}\,\mathbf{a} - \frac{M}{\sqrt{EG}}\,\mathbf{b} + \frac{G_1}{2G\sqrt{E}}\,\mathbf{n}.$

If now we form the vector product curl **a** × curl **b**, the coefficient of **n** in the expression is equal to $(LN - M^2)/H^2$ or K. Hence the second curvature is given by

$$K = \mathbf{n} \cdot \mathrm{curl}\,\mathbf{a} \times \mathrm{curl}\,\mathbf{b}$$
$$= [\mathbf{n},\ \mathrm{curl}\,\mathbf{a},\ \mathrm{curl}\,\mathbf{b}] \quad \dots\dots\dots\dots(23).$$

Ex. Deduce from (22) that

$$\kappa_g = \mathbf{n} \bullet \operatorname{curl} (\mathbf{r}_1 u' + \mathbf{r}_2 v')$$

$$= \frac{1}{H} \frac{\partial}{\partial u} (Fu' + Gv') - \frac{1}{H} \frac{\partial}{\partial v} (Eu' + Fv').$$

EXAMPLES XV

1. Prove that $\operatorname{div} \mathbf{r} = 2$ and $\operatorname{curl} \mathbf{r} = 0$.

2. Verify the values of curl $P\mathbf{r}_1$ and curl $Q\mathbf{r}_2$ given in Art. 118.

3. If ϕ is a point-function, and F is a function of ϕ, show that

$$\nabla F = F' \nabla \phi.$$

Hence, by means of (17), prove that

$$\nabla^2 F = F'' (\nabla \phi)^2 + F' \nabla^2 \phi.$$

4. If $F - F(\phi, \psi, ...)$ is a function of several point-functions, show that

$$\nabla F = \frac{\partial F}{\partial \phi} \nabla \phi + \frac{\partial F}{\partial \psi} \nabla \psi + ...$$

5. If u, v are geodesic polar coordinates, so that $E = 1$, $F = 0$ and $H^2 = G$, show that

$$\nabla \phi = \phi_1 \mathbf{r}_1 + \frac{1}{G} \phi_2 \mathbf{r}_2$$

and $$\nabla^2 \phi = \frac{1}{H} \frac{\partial}{\partial u} (H\phi_1) + \frac{1}{H} \frac{\partial}{\partial v} \left(\frac{\phi_2}{H} \right).$$

Hence, if H is a function of u only, show that $\int \frac{du}{H}$ satisfies the equation $\nabla^2 \phi = 0$; and also that

$$K = - \nabla^2 \log H.$$

6. If $\mathbf{t}$ is a unit vector tangential to the surface, and $\mathbf{b} = \mathbf{t} \times \mathbf{n}$, show that the normal curvature in the direction of $\mathbf{t}$ is $-(\mathbf{t} \bullet \nabla \mathbf{n}) \bullet \mathbf{t}$, and the torsion of the geodesic in this direction $(\mathbf{t} \bullet \nabla \mathbf{n}) \bullet \mathbf{b}$. Deduce Euler's theorem on normal curvature (Art. 31), and the formula $(\kappa_b - \kappa_a) \sin \theta \cos \theta$ for the torsion of the geodesic.

7. Show that the directions of $\mathbf{c}$ and $\mathbf{d}$ on a surface are conjugate if $(\mathbf{c} \bullet \nabla \mathbf{n}) \bullet \mathbf{d} = 0$; and hence that the asymptotic directions are such that $(\mathbf{d} \bullet \nabla \mathbf{n}) \bullet \mathbf{d} = 0$. Deduce the differential equation of the asymptotic lines

$$(d\mathbf{r} \bullet \nabla \mathbf{n}) \bullet d\mathbf{r} = 0.$$

8. If $\mathbf{t}$ is the unit tangent to a line of curvature, show that $(\mathbf{t} \bullet \nabla \mathbf{n}) \times \mathbf{t} = 0$. Deduce the differential equation of the lines of curvature

$$(d\mathbf{r} \bullet \nabla \mathbf{n}) \times d\mathbf{r} = 0.$$

9. If $\mathbf{a}$, $\mathbf{b}$ are the unit tangents to the orthogonal parametric curves (Art. 121), show that

$$K = (\mathbf{a} \bullet \nabla \mathbf{a}) \bullet (\mathbf{b} \bullet \nabla \mathbf{b}) - (\mathbf{b} \bullet \nabla \mathbf{a}) \bullet (\mathbf{a} \bullet \nabla \mathbf{b}),$$

and also that $K = - \operatorname{div} (\mathbf{a} \operatorname{div} \mathbf{a} + \mathbf{b} \operatorname{div} \mathbf{b})$.

10. As in Art. 120, prove the formulae

$$\nabla(\phi\psi) = \phi\nabla\psi + \psi\nabla\phi,$$

$$\nabla\times(\mathbf{U}\times\mathbf{V}) = \mathbf{V}\cdot\nabla\mathbf{U} - \mathbf{U}\cdot\nabla\mathbf{V} + \mathbf{U}\nabla\cdot\mathbf{V} - \mathbf{V}\nabla\cdot\mathbf{U},$$

$$\nabla(\mathbf{U}\cdot\mathbf{V}) = \mathbf{V}\cdot\nabla\mathbf{U} + \mathbf{U}\cdot\nabla\mathbf{V} + \mathbf{V}\times\nabla\times\mathbf{U} + \mathbf{U}\times\nabla\times\mathbf{V}.$$

11. If $\mathbf{c}$ is a constant vector, show that

$$\nabla(\mathbf{c}\cdot\mathbf{U}) = \mathbf{c}\cdot\nabla\mathbf{U} + \mathbf{c}\times\operatorname{curl}\mathbf{U},$$

$$\nabla\cdot(\mathbf{c}\times\mathbf{U}) = -\mathbf{c}\cdot\operatorname{curl}\mathbf{U},$$

$$\nabla\times(\mathbf{c}\times\mathbf{U}) = \mathbf{c}\operatorname{div}\mathbf{U} - \mathbf{c}\cdot\nabla\mathbf{U}.$$

12. If $\mathbf{a}$ is tangential to the surface, show that

$$\mathbf{a}\cdot\nabla\mathbf{r} = \mathbf{a}.$$

And, if $\mathbf{c}$ is a constant vector,

$$\nabla(\mathbf{c}\cdot\mathbf{r}) = \mathbf{c}\cdot\nabla\mathbf{r}, \quad \nabla\cdot(\mathbf{c}\times\mathbf{r}) = 0,$$

$$\nabla\times(\mathbf{c}\times\mathbf{r}) = 2\mathbf{c} - \mathbf{c}\cdot\nabla\mathbf{r}.$$

13. Prove that $\quad\mathbf{U}\cdot\nabla\mathbf{U} = \tfrac{1}{2}\nabla\mathbf{U}^2 - \mathbf{U}\times\operatorname{curl}\mathbf{U}.$

14. If $\mathbf{r}$ is the position vector of the current point on the surface, and $p = \mathbf{r}\cdot\mathbf{n}$, prove that $\nabla\mathbf{r}^2$ is twice the tangential component of $\mathbf{r}$, and that

$$(\nabla\mathbf{r}^2)^2 = 4(\mathbf{r}^2 - p^2).$$

Also show that $\qquad\nabla^2\mathbf{r}^2 = 2(2 + pJ).$

15. If F is a function of ϕ and ψ, deduce from Ex. 4 that

$$\nabla^2 F = \frac{\partial F}{\partial\phi}\nabla^2\phi + \frac{\partial F}{\partial\psi}\nabla^2\psi + \frac{\partial^2 F}{\partial\phi^2}(\nabla\phi)^2 + 2\frac{\partial^2 F}{\partial\phi\partial\psi}\nabla\phi\cdot\nabla\psi + \frac{\partial^2 F}{\partial\psi^2}(\nabla\psi)^2$$

and that

$$(\nabla F)^2 = \left(\frac{\partial F}{\partial\phi}\right)^2(\nabla\phi)^2 + 2\frac{\partial F}{\partial\phi}\frac{\partial F}{\partial\psi}\nabla\phi\cdot\nabla\psi + \left(\frac{\partial F}{\partial\psi}\right)^2(\nabla\psi)^2.$$

16. If x, y, z are the rectangular coordinates of the current point on the surface, and l, m, n the direction cosines of the normal, show that

$$(\nabla x)^2 + (\nabla y)^2 + (\nabla z)^2 = 2$$

and $\qquad(\nabla l)^2 + (\nabla m)^2 + (\nabla n)^2 = J^2 - 2K.$

17. If $\gamma = |\nabla\phi|$, prove that the geodesic curvature of the curve $\phi = \text{const.}$ is given by

$$\kappa_g = \frac{\gamma\nabla^2\phi - \nabla\gamma\cdot\nabla\phi}{(\nabla\phi)^2}.$$

18. Prove that a family of geodesics is characterised by the property $\mathbf{n}\cdot\operatorname{curl}\mathbf{t} = 0$, $\mathbf{t}$ being the unit tangent. Deduce that $\mathbf{t}$ is the gradient of some scalar function ψ; and that the curves $\psi = \text{const.}$ are the geodesic parallels to the family of geodesics, ψ measuring the actual geodesic distance from a fixed parallel.

19. Show that the equation of the indicatrix at a point is $(\mathbf{r}\cdot\nabla\mathbf{n})\cdot\mathbf{r} = -1$, the point itself being the origin of position vectors.

20. Prove that the second curvature is given by the formula

$$2K = (\nabla^2 \mathbf{r})^2 - \underset{x,\,y,\,z}{\Sigma} (\text{curl grad } x)^2.$$

21. If $p = \mathbf{r} \cdot \mathbf{n}$, show that $\quad \nabla p = \mathbf{r} \cdot \nabla \mathbf{n}$

and

$$\nabla^2 p = p(2K - J^2) - J - \mathbf{r} \cdot \nabla J$$
$$= p(2K - J^2) + J - \text{div } J\mathbf{r}.$$

Hence, in the case of a minimal surface,

$$\nabla^2 p = 2pK.$$

22. Prove the relations

$$\text{div curl } \phi \mathbf{V} = \mathbf{V} \cdot \nabla \times \nabla \phi + \phi \nabla \cdot \nabla \times \mathbf{V},$$
$$\nabla^2 (\phi \mathbf{V}) = \phi \nabla^2 \mathbf{V} + 2 \nabla \phi \cdot \nabla \mathbf{V} + \mathbf{V} \nabla^2 \phi.$$

TRANSFORMATION OF INTEGRALS

122. Divergence theorem. We shall now prove various
theorems connecting line integrals round a closed curve drawn on
the surface, with surface integrals over the enclosed region. These
are analogous to the three-parametric theorems of Gauss, Stokes
and Green, and others deducible from them. Let C be any closed
curve drawn on the surface; and at any point of this curve let $\mathbf{m}$
be the unit vector tangential to the surface and normal to the
curve, drawn *outward* from the region enclosed by C. Let $\mathbf{t}$ be the
unit tangent to the curve, in that sense for which $\mathbf{m}$, $\mathbf{t}$, $\mathbf{n}$ form
a right-handed system of unit vectors, so that

$$\mathbf{m} = \mathbf{t} \times \mathbf{n}, \quad \mathbf{t} = \mathbf{n} \times \mathbf{m}, \quad \mathbf{n} = \mathbf{m} \times \mathbf{t}.$$

The sense of $\mathbf{t}$ is the positive sense for a description of the curve.
If ds is the length of an element of the curve, the corresponding

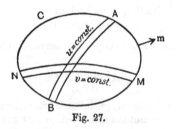

Fig. 27.

displacement $d\mathbf{r}$ along the curve in the positive sense is given by
$d\mathbf{r} = \mathbf{t}\,ds$.

Consider first a transformation of the surface integral of the
divergence of a vector over the region enclosed by C. The area

dS of an element of this region is equal to $H\,du\,dv$. If the vector
F is given by

$$\mathbf{F} = P\mathbf{r}_1 + Q\mathbf{r}_2 + R\mathbf{n},$$

then by (7)

$$\text{div } \mathbf{F} = \frac{1}{H}\left[\frac{\partial}{\partial u}(HP) + \frac{\partial}{\partial v}(HQ)\right] - JR$$

and the definite integral of div **F** over the portion of the surface
enclosed by C is

$$\iint \text{div } \mathbf{F}\,dS = \iint \left[\frac{\partial}{\partial u}(HP) + \frac{\partial}{\partial v}(HQ)\right]du\,dv - \iint JR\,dS$$

$$= \int [HP]_N^M\,dv + \int [HQ]_B^A\,du - \iint JR\,dS,$$

N, M being the points in which the curve $v = $ const. meets C, and
B, A those in which $u = $ const. meets it. If now we assign to du at
the points B, A and to dv at the points N, M the values corresponding to the passage round C in the positive sense, the above
equation becomes

$$\iint \text{div } \mathbf{F}\,dS = \int_0 HP\,dv - \int_0 HQ\,du - \iint JR\,dS.$$

Consider now the line integral $\int_0 \mathbf{F} \cdot \mathbf{m}\,ds$ taken round C in the
positive sense. Clearly $R\mathbf{n} \cdot \mathbf{m} = 0$, and $\mathbf{m}\,ds = \mathbf{t} \times \mathbf{n}\,ds = d\mathbf{r} \times \mathbf{n}$.
Hence

$$\int_0 \mathbf{F} \cdot \mathbf{m}\,ds = \int_0 (P\mathbf{r}_1 + Q\mathbf{r}_2) \cdot (\mathbf{r}_1\,du + \mathbf{r}_2\,dv) \times \frac{(\mathbf{r}_1 \times \mathbf{r}_2)}{H}.$$

In the integrand the coefficient of du is

$$\frac{1}{H}(P\mathbf{r}_1 + Q\mathbf{r}_2) \cdot (F\mathbf{r}_1 - E\mathbf{r}_2) = -HQ$$

and the coefficient of dv is

$$\frac{1}{H}(P\mathbf{r}_1 + Q\mathbf{r}_2) \cdot (G\mathbf{r}_1 - F\mathbf{r}_2) = HP,$$

so that

$$\int_0 \mathbf{F} \cdot \mathbf{m}\,ds = \int_0 HP\,dv - \int_0 HQ\,du.$$

Comparing this with the value found for the surface integral of the
divergence, we have the required result, which may be written

$$\iint \text{div } \mathbf{F}\,dS = \int_0 \mathbf{F} \cdot \mathbf{m}\,ds - \iint J\mathbf{F} \cdot \mathbf{n}\,dS \quad\ldots\ldots\ldots(24).$$

This is analogous to Gauss's "divergence theorem," and we shall therefore refer to it as the *divergence theorem*. The last term in (24) has no counterpart in Gauss's theorem, but it has some important consequences in geometry of surfaces, and in physical problems connected therewith.

From this theorem the *invariant property* of div $\mathbf{F}$ follows immediately. For, by letting the curve C converge to a point P inside it, we have for that point

$$\text{div } \mathbf{F} + J\mathbf{n} \cdot \mathbf{F} = \text{Lt} \frac{\displaystyle\int_o \mathbf{F} \cdot \mathbf{m} \, ds}{dS} \quad \ldots\ldots\ldots\ldots(25).$$

Now the second member of this equation, and also the second term of the first member, are clearly independent of the choice of coordinates. Hence div $\mathbf{F}$ must also be independent of it, and is thus an invariant. This equation may also be regarded as giving an alternative definition of div $\mathbf{F}$.

123. Other theorems. From the divergence theorem other important transformations are easily deducible. If, for instance, in (24) we put $\mathbf{F} = \phi\mathbf{c}$, where ϕ is a scalar function and $\mathbf{c}$ a constant vector, we find in virtue of (17),

$$\iint \nabla\phi \cdot \mathbf{c} \, dS = \int_o \phi\mathbf{c} \cdot \mathbf{m} \, ds - \iint J\phi\mathbf{c} \cdot \mathbf{n} \, dS.$$

And, since this is true for all values of the constant vector $\mathbf{c}$, it follows that

$$\iint \nabla\phi \, dS = \int_o \phi\mathbf{m} \, ds - \iint J\phi\mathbf{n} \, dS \quad \ldots\ldots\ldots\ldots(26).$$

This theorem has some important applications, both geometrical and physical. Putting ϕ equal to a constant we obtain the formula

$$\int_o \mathbf{m} \, ds = \iint J\mathbf{n} \, dS \quad \ldots\ldots\ldots\ldots\ldots\ldots(27).$$

If now we let the curve C converge to a point inside it, the last equation gives

$$J\mathbf{n} = \text{Lt} \frac{\displaystyle\int_o \mathbf{m} \, ds}{dS} \quad \ldots\ldots\ldots\ldots\ldots\ldots(28).$$

Hence we have an alternative definition of the *first curvature* of a surface, independent of normal curvature or principal directions. We may state it:

The limiting value of the line integral $\int_o \mathbf{m}\, ds$, *per unit of enclosed area, is normal to the surface, and its ratio to the unit normal is equal to the first curvature.*

In the case of a closed surface another important result follows from (27). For we may then let the curve C converge to a point *outside* it. The line integral in (27) then tends to zero, and the surface integral over the whole surface must vanish. Thus, *for a closed surface,*

$$\iint J\mathbf{n}\, dS = 0,$$

the integral being taken over the whole surface. In virtue of (13) we may also write this

$$\iint \nabla^2 \mathbf{r}\, dS = 0.$$

Again, apply the divergence theorem to the vector $\mathbf{F} \times \mathbf{c}$, where $\mathbf{c}$ is a constant vector. Then by (19) the theorem becomes

$$\mathbf{c} \cdot \iint \operatorname{curl} \mathbf{F}\, dS = \mathbf{c} \cdot \int_o \mathbf{m} \times \mathbf{F}\, ds - \mathbf{c} \cdot \iint J\mathbf{n} \times \mathbf{F}\, dS.$$

And, since this is true for all values of the constant vector $\mathbf{c}$, we have

$$\iint \operatorname{curl} \mathbf{F}\, dS = \int_o \mathbf{m} \times \mathbf{F}\, ds - \iint J\mathbf{n} \times \mathbf{F}\, dS \quad\ldots\ldots(29).$$

This important result may be used to prove the *invariant property* of curl $\mathbf{F}$. For, on letting the curve C converge to a point inside it, we have at this point

$$\operatorname{curl} \mathbf{F} + J\mathbf{n} \times \mathbf{F} = \operatorname{Lt} \frac{\int_o \mathbf{m} \times \mathbf{F}\, ds}{dS} \quad\ldots\ldots\ldots(30).$$

Now each term of this equation, except curl $\mathbf{F}$, is independent of the choice of coordinates. Hence curl $\mathbf{F}$ must also be independent. It is therefore an invariant. The equation (30) may be regarded as giving an alternative definition of curl $\mathbf{F}$.

In the case of a *minimal surface*, $J = 0$. Thus (26) becomes

$$\iint \nabla \phi\, dS = \int_o \phi\, \mathbf{m}\, ds,$$

and from (27) we see that

$$\int_o \mathbf{m}\, ds = 0$$

for any closed curve drawn on the surface. Similarly (29) becomes

$$\iint \operatorname{curl} \mathbf{F}\, dS = \int_o \mathbf{m} \times \mathbf{F}\, ds.$$

In particular if we put for $\mathbf{F}$ the position vector $\mathbf{r}$ of the current point, since curl $\mathbf{r} = 0$ we obtain

$$\int_o \mathbf{r} \times \mathbf{m}\, ds = 0.$$

This equation and the equation $\int_o \mathbf{m}\, ds = 0$ are virtually the equations of equilibrium of a thin film of constant tension, with equal pressures on the two sides. The one equation expresses that the vector sum of the forces on the portion enclosed by C is zero, the other that the vector sum of their moments about the origin vanishes *.

Analogues of Green's theorems are easily deducible from the divergence theorem. For if we apply this theorem to the function $\phi \nabla \psi$, which is tangential to the surface, since by (17)

$$\operatorname{div}(\phi \nabla \psi) = \nabla \phi \cdot \nabla \psi + \phi \nabla^2 \psi$$

the divergence theorem gives

$$\int_o \phi \nabla \psi \cdot \mathbf{m}\, ds = \iint (\nabla \phi \cdot \nabla \psi + \phi \nabla^2 \psi)\, dS.$$

Transposing terms we may write this

$$\iint \nabla \phi \cdot \nabla \psi\, dS = \int_o \phi \mathbf{m} \cdot \nabla \psi\, ds - \iint \phi \nabla^2 \psi\, dS \dots\dots(31).$$

On interchanging ϕ and ψ we have similarly

$$\iint \nabla \phi \cdot \nabla \psi\, dS = \int_o \psi \mathbf{m} \cdot \nabla \phi\, ds - \iint \psi \nabla^2 \phi\, dS \dots\dots(32).$$

These have the same form as the well-known theorems due to Green. From (31) and (32) we also have the symmetrical relation

$$\int_o (\phi \nabla \psi - \psi \nabla \phi) \cdot \mathbf{m}\, ds = \iint (\phi \nabla^2 \psi - \psi \nabla^2 \phi)\, dS \quad \dots(33).$$

* For the application of these theorems to the equilibrium of stretched membranes, and the flow of heat in a curved lamina, see the author's paper already referred to, §§ 15–17.

If in this formula we put $\psi = $ const., we obtain the theorem

$$\int_{0} \mathbf{m} \cdot \nabla \phi \, ds = \iint \nabla^2 \phi \, dS \quad .... \quad ....(34),$$

which could also be deduced from the divergence theorem by putting $\mathbf{F} = \nabla \phi$.

Geodesic polar coordinates and the concept of geodesic distance may be used to extend this theory in various directions*. But for fear of overloading the present chapter we shall refrain from doing this.

124. Circulation theorem. Consider next the definite integral of $\mathbf{n} \cdot \operatorname{curl} \mathbf{F}$ over the portion of the surface enclosed by the curve C (Fig. 27). If, as before, $\mathbf{F} = P\mathbf{r}_1 + Q\mathbf{r}_2 + R\mathbf{n}$, we have in virtue of (12)

$$\mathbf{n} \cdot \operatorname{curl} \mathbf{F} = \frac{1}{H} \left\{ \frac{\partial}{\partial u} (FP + GQ) - \frac{\partial}{\partial v} (EP + FQ) \right\}.$$

Hence, since $dS = H \, du \, dv$, the definite integral referred to is

$$\iint \mathbf{n} \cdot \nabla \times \mathbf{F} \, dS = \iint \left\{ \frac{\partial}{\partial u} (FP + GQ) - \frac{\partial}{\partial v} (EP + FQ) \right\} \, du \, dv$$

$$= \int [FP + GQ]_{N}^{M} dv - \int [EP + FQ]_{B}^{A} du.$$

If now we assign to dv at N, M and to du at B, A the values corresponding to the passage round C in the positive sense, this becomes

$$\iint \mathbf{n} \cdot \nabla \times \mathbf{F} \, dS = \int_{0} (FP + GQ) \, dv + \int_{0} (EP + FQ) \, du.$$

But the line integral

$$\int_{0} \mathbf{F} \cdot d\mathbf{r} = \int_{0} (P\mathbf{r}_1 + Q\mathbf{r}_2 + R\mathbf{n}) \cdot (\mathbf{r}_1 du + \mathbf{r}_2 dv)$$

$$= \int_{0} (EP + FQ) \, du + \int_{0} (FP + GQ) \, dv.$$

The two integrals are therefore equal, so that

$$\iint \mathbf{n} \cdot \operatorname{curl} \mathbf{F} \, dS = \int_{0} \mathbf{F} \cdot d\mathbf{r} \quad ....(35).$$

This may be referred to as the *circulation theorem,* and the integral in the second member as the *circulation* of the vector $\mathbf{F}$ round the curve C. The theorem is analogous to Stokes's theorem, and is

* *Loc. cit.,* § 14.

virtually identical with it, since the normal resolute of our function curl $\mathbf{F}$ is equal to the normal resolute of the three-parametric function.

If we apply the above theorem to the function $\phi\mathbf{c}$, where ϕ is a scalar function and $\mathbf{c}$ a constant vector, we find in virtue of (18)

$$\iint \mathbf{n} \cdot \nabla\phi \times \mathbf{c}\, dS = \int_0 \phi\mathbf{c} \cdot d\mathbf{r}.$$

And, since this is true for all values of the constant vector $\mathbf{c}$, we have the theorem

$$\iint \mathbf{n} \times \nabla\phi\, dS = \int_0 \phi\, d\mathbf{r} \quad \dots\dots\dots\dots\dots(36),$$

which is sometimes useful.

If we apply the circulation theorem to the function $\nabla\phi$, we find

$$\iint \mathbf{n} \cdot \nabla \times \nabla\phi\, dS = \int_0 \nabla\phi \cdot d\mathbf{r} = 0.$$

And since this is true for the region bounded by any closed curve, it follows that $\mathbf{n} \cdot \text{curl grad}\, \phi$ vanishes identically. Thus *the curl of the gradient of a scalar function is tangential to the surface*, as already proved in Art. 118. Conversely, suppose that both $\mathbf{F}$ and curl $\mathbf{F}$ are tangential to the surface. Then, by the circulation theorem,

$$\int_0 \mathbf{F} \cdot d\mathbf{r} = \iint \mathbf{n} \cdot \text{curl } \mathbf{F}\, dS = 0.$$

And, since this is true for any closed curve, it follows from Art. 115 that $\mathbf{F}$ is the gradient of some scalar function. Hence the theorem:

If a vector function and its curl are both tangential to the surface, the vector is the gradient of some scalar function.

EXAMPLES XVI

1. Show that $\int_0 \mathbf{r} \cdot d\mathbf{r} = 0$ is true for any closed curve.

2. By applying the divergence theorem to the function curl $R\mathbf{n}$, prove that div curl $R\mathbf{n}$ vanishes identically.

3. For a *closed* surface, prove that the integrals

$$\iint \mathbf{n} \cdot \text{curl } \mathbf{F}\, dS, \quad \iint \mathbf{n} \times \nabla\phi\, dS$$

vanish identically.

4. Prove the relations

$$\iint \mathbf{V} \cdot \nabla\phi \, dS = \int_{\circ} \phi \mathbf{V} \cdot \mathbf{m} \, ds - \iint (\phi \operatorname{div} \mathbf{V} + J\phi \mathbf{n} \cdot \mathbf{V}) \, dS$$

and

$$\iint \mathbf{V} \cdot \nabla \times \mathbf{U} \, dS = \int_{\circ} [\mathbf{U}, \mathbf{V}, \mathbf{m}] \, ds + \iint (\mathbf{U} \cdot \nabla \times \mathbf{V} - J\mathbf{U} \times \mathbf{V} \cdot \mathbf{n}) \, dS.$$

5. If $\mathbf{V} = \nabla\phi$ and $\nabla^2\phi = 0$, show that

$$\iint \mathbf{V}^2 \, dS = \int_{\circ} \phi \mathbf{m} \cdot \mathbf{V} \, ds.$$

6. If $\mathbf{F} = \nabla\phi$ and $\nabla^2\phi = -2\pi\sigma$, show that

$$\int_{\circ} \mathbf{F} \cdot \mathbf{m} \, ds = \iint -2\pi\sigma \, dS.$$

7. Show that $\qquad \int_{\circ} \phi\nabla\psi \cdot d\mathbf{r} = -\int_{\circ} \psi\nabla\phi \cdot d\mathbf{r}.$

8. From (18) and the circulation theorem deduce the relation

$$\iint \phi \mathbf{n} \cdot \nabla \times \mathbf{V} \, dS = \int_{\circ} \phi \mathbf{V} \cdot d\mathbf{r} - \iint \nabla\phi \times \mathbf{V} \cdot \mathbf{n} \, dS.$$

Putting $\mathbf{V} = \nabla\psi$ in this result, prove that

$$\iint \nabla\phi \times \nabla\psi \cdot \mathbf{n} \, dS = \int_{\circ} \phi\nabla\psi \cdot d\mathbf{r} = -\int_{\circ} \psi\nabla\phi \cdot d\mathbf{r}.$$

9. If $c\rho \dfrac{\partial\phi}{\partial t} = \operatorname{div}(\mu\nabla\phi)$, and ϕ vanishes over the closed curve C, show that

$$\iint c\rho\phi \frac{\partial\phi}{\partial t} \, dS = -\iint \mu(\nabla\phi)^2 \, dS.$$

10. If $p = \mathbf{r} \cdot \mathbf{n}$ and S is the area of the surface bounded by the closed curve C, show that

$$2S = \int_{\circ} \mathbf{r} \cdot \mathbf{m} \, ds - \iint Jp \, dS.$$

11. Prove the relation

$$\iint \nabla J \, dS = \int_{\circ} J\mathbf{m} \, ds - \iint J^2 \mathbf{n} \, dS.$$

Hence deduce that $\qquad 2K\mathbf{n} = \nabla^2\mathbf{n} + \operatorname{Lt} \dfrac{\displaystyle\int_{\circ} J\mathbf{m} \, ds}{dS}.$

12. Show that, for any closed curve,

$$\int_{\circ} \mathbf{m} \times \mathbf{r} \, ds = \iint J\mathbf{n} \times \mathbf{r} \, dS.$$

13. Deduce formula (34) from the divergence theorem.

14. The pole for geodesic polar coordinates is inside the closed curve C, and u is the geodesic distance from the pole. In formula (33) put $\phi = \log u$, isolating the pole with a small geodesic circle. Letting this geodesic circle converge to the pole, deduce the *analogue of Green's formula* for the value of ψ at the pole, viz.

$$2\pi\psi = \int_\bullet (\psi\nabla\log u - \log u\nabla\psi)\bullet \mathbf{m}\,ds + \iint(\log u\nabla^2\psi - \psi\nabla^2\log u)\,dS.$$

Hence show that $2\pi = \int_\bullet \mathbf{m}\bullet\nabla\log u\,ds - \iint\nabla^2\log u\,dS.$

15. Show that the formulae of Ex. 14 are true with $\log H$ in place of $\log u$.

16. If, in Ex. 14, H is a function of u only, the function

$$\Omega = \int\frac{du}{H}$$

satisfies $\nabla^2\Omega = 0$. Using Ω in place of $\log u$ in that exercise, prove the formulae

$$2\pi\psi = \int_\bullet (\psi\nabla\Omega - \Omega\nabla\psi)\bullet\mathbf{m}\,ds + \iint\Omega\nabla^2\psi\,dS$$

and $2\pi = \int_\bullet \mathbf{m}\bullet\nabla\Omega\,ds.$

The latter is analogous to Gauss's integral for 4π.

17 Prove the following generalisation of (31):

$$\iint W\nabla U\bullet\nabla V\,dS = \int_\bullet UW\nabla V\bullet\mathbf{m}\,ds - \iint U\operatorname{div}(W\nabla V)\,dS$$
$$= \int_\bullet VW\nabla U\bullet\mathbf{m}\,ds - \iint V\operatorname{div}(W\nabla U)\,dS.$$

18. *A necessary and sufficient condition that a family of curves on a surface be parallels is that the divergence of the unit tangent vanish identically.* (See Art. 130.)

19. *The orthogonal trajectories of a family of parallels constitute a family of geodesics; and conversely.* (Ex. 18.)

20. The surface integral of the geodesic curvature of a family of curves over any region is equal to the circulation of the unit tangent round the boundary of the region. Hence this circulation vanishes for a family of geodesics.

21. *If $\mathbf{s}$ is a vector point-function for a given surface, the vector $\mathbf{s}_1\times\mathbf{s}_2/H$ is independent of the choice of parametric curves.* (Art. 131.)

22. *A necessary and sufficient condition that an orthogonal system of curves on a surface may be isometric is that, at any point, the sum of the derivatives of the geodesic curvatures of the curves, each in its own direction, be zero.*

23. *An orthogonal system of curves cutting an isometric orthogonal system at a variable angle θ will itself be isometric provided $\nabla^2\theta = 0$.*

CONCLUSION

FURTHER RECENT ADVANCES

125. Orthogonal systems of curves on a surface. Since this book was sent to the press, several important additions have been made by the author to our knowledge of the properties of families of curves and surfaces, and of the general small deformation of a surface. A brief account is here given of the new results established; and it will be seen that the two-parametric divergence and curl introduced in Chapter XII play an important part in the theory.

We are already familiar with the theorem of Dupin, which states that the sum of the normal curvatures of a surface in any two perpendicular directions at a point is invariant, and equal to the first curvature of the surface (Art. 31). The author has shown that this is only one aspect of a more comprehensive theorem dealing with the curvature of orthogonal systems of curves drawn on the surface—a theorem specifying both the first and the second curvatures[*]. Let the orthogonal system considered be taken as parametric curves, and let $\mathbf{a}$, $\mathbf{b}$ be the unit vectors tangential to these curves. Then the vector curvature of the curve $v = \text{const.}$ is (Art. 52)

$$\frac{1}{\sqrt{E}} \frac{\partial \mathbf{a}}{\partial u} = \frac{L}{E} \mathbf{n} - \frac{E_2}{2E\sqrt{G}} \mathbf{b} = \frac{L}{E} \mathbf{n} - \mathbf{b} \operatorname{div} \mathbf{b}$$

and that of the curve $u = \text{const.}$ is

$$\frac{1}{\sqrt{G}} \frac{\partial \mathbf{b}}{\partial v} = \frac{N}{G} \mathbf{n} - \frac{G_1}{2G\sqrt{E}} \mathbf{a} = \frac{N}{G} \mathbf{n} - \mathbf{a} \operatorname{div} \mathbf{a}.$$

The sum of these vector curvatures has a component $J\mathbf{n}$ normal to the surface, and a component $-(\mathbf{a} \operatorname{div} \mathbf{a} + \mathbf{b} \operatorname{div} \mathbf{b})$ tangential to the surface. Now the divergence of the latter component, by (7) of Art. 116, has the value

$$-\frac{1}{2\sqrt{EG}} \left[\frac{\partial}{\partial u} \left(\frac{G_1}{\sqrt{EG}} \right) + \frac{\partial}{\partial v} \left(\frac{E_2}{\sqrt{EG}} \right) \right]$$

$$= -\frac{1}{EG} \left[\frac{\partial}{\partial u} \left(\frac{1}{\sqrt{E}} \frac{\partial \sqrt{G}}{\partial u} \right) + \frac{\partial}{\partial v} \left(\frac{1}{\sqrt{G}} \frac{\partial \sqrt{E}}{\partial v} \right) \right],$$

[*] "Some New Theorems in Geometry of a Surface," *The Mathematical Gazette*, Vol. 13, pp. 1—6 (January 1926).

which is equal to the second curvature K, in virtue of the Gauss characteristic equation. Hence the theorem:

The sum of the vector curvatures of the two curves of an orthogonal system through any point has a normal component whose magnitude is equal to the first curvature of the surface, and a tangential component whose divergence is equal to the second curvature at that point.

The normal component of this vector curvature is thus invariant, being the same for all orthogonal systems. This is substantially Dupin's theorem. The tangential component is not itself invariant; but it possesses an invariant divergence. The behaviour of this component is expressed by the following theorem*:

The vector curvature is the same for orthogonal systems that cut each other at a constant angle. If, however, the inclination θ of one system to the other is variable, their curvatures differ by the tangential vector curl $(\theta \mathbf{n})$, whose divergence vanishes identically.

Since the divergence of the normal component $J\mathbf{n}$ is equal to $-J^2$, we also have the result:

The divergence of the vector curvature of an orthogonal system on the surface is invariant and equal to $K - J^2$.

126. Family of curves on a surface. Again the author has shown† that many of the properties possessed by the generators of a ruled surface do not belong exclusively to families of straight lines on a surface; but that a family of curves on any surface possesses a *line of striction* and a *focal curve* or *envelope*, though these are not necessarily real. When the surface is developable, and the curves are the generators, the focal curve is the edge of regression.

Consider then a singly infinite family of curves on a given surface, and let these be taken as the parametric curves $v = $ const. If a curve v meets a consecutive curve $v + dv$, a point on the former corresponding to parameter values (u, v) must be identical with some point on the latter with parameter values $(u + du, v + dv)$; or

$$\mathbf{r}(u, v) = \mathbf{r}(u + du, v + dv).$$

---

* For the proof of this theorem see § 6 of the author's paper just referred to.
† *Loc. cit.*, §§ 1—4.

Hence $\qquad\qquad\mathbf{r}_1 du + \mathbf{r}_2 dv = 0.$

Now this is possible only where $\mathbf{r}_1$ is parallel to $\mathbf{r}_2$; that is to say where
$$\mathbf{r}_1 \times \mathbf{r}_2 = H\mathbf{n} = 0.$$
Hence the required locus of points of intersection of consecutive curves of the family is given by
$$H = 0.$$
This may be called the *focal curve* or *envelope* of the family; and the points in which it is met by any curve are the *foci* of that curve. A curve touches the focal curve at each of its foci. Also this focal curve is clearly the focal curve of the family $u = \text{const.}$, and of the family $\phi(u, v) = \text{const.}$

Consider next the possibility of a normal to a curve of the family $v = \text{const.}$ being normal also to a consecutive curve. The author has shown [*] that this is possible only where
$$\frac{\partial}{\partial u}\left(\frac{H^2}{E}\right) = 0, \quad \text{that is} \quad \frac{H}{\sqrt{E}}\frac{\partial}{\partial u}\left(\frac{H}{\sqrt{E}}\right) = 0.$$
Now $H = 0$ is the equation of the focal curve. Hence the locus of points possessing the required property is
$$\frac{\partial}{\partial u}\left(\frac{H}{\sqrt{E}}\right) = 0, \quad \text{or} \quad \text{div } \mathbf{a} = 0.$$
This locus may be called the *line of striction* of the family of curves, the line of striction of the generators of a ruled surface being a particular case. Hence the theorem:

Given a one-parameter family of curves on a surface, with **t** *as the unit tangent, the equation of the line of striction of the family may be expressed*
$$\text{div } \mathbf{t} = 0.$$
Or, since div **t** is the geodesic curvature of the orthogonal trajectories of the family of curves, we have the result:

The line of striction of a family of curves is the locus of points at which the geodesic curvature of their orthogonal trajectories is zero.

An important example is that of a family of geodesics; and for such a family the author has proved [†] the following extension of Bonnet's theorem on the generators of a ruled surface (Art. 72):

---

[*] *Loc. cit.*, §2.   [†] *Loc. cit.*, §3.

If a curve is drawn on a surface so as to cut a family of geodesics, then provided it has two of the following properties it will also have the third: (a) that it is a geodesic, (b) that it is the line of striction of the family of geodesics, (c) that it cuts the family at a constant angle.

Another theorem is connected with the curl of the unit tangent to the family of curves. Since

$$\text{curl } \mathbf{a} = \frac{M}{\sqrt{EG}} \mathbf{a} - \frac{L}{E} \mathbf{b} - \frac{E_2}{2E\sqrt{G}} \mathbf{n},$$

it follows that $\mathbf{a} \cdot \text{curl } \mathbf{a}$ is equal to the torsion of the geodesic tangent to the curve $v = \text{const.}$ (Art. 49), $\mathbf{n} \cdot \text{curl } \mathbf{a}$ to the geodesic curvature of the curve (Art. 54), and $-\mathbf{b} \cdot \text{curl } \mathbf{a}$ to the normal curvature of the surface in its direction. Hence the theorem:

If $\mathbf{t}$ is the unit tangent at any point to the curve of a family, the geodesic curvature of the curve is $\mathbf{n} \cdot \text{curl } \mathbf{t}$, the torsion of its geodesic tangent is $\mathbf{t} \cdot \text{curl } \mathbf{t}$, and the normal curvature of the surface in the direction of the curve is $\mathbf{t} \times \mathbf{n} \cdot \text{curl } \mathbf{t}$.

Since these three quantities vanish for geodesics, lines of curvature and asymptotic lines respectively, it follows that:

A family of curves with a unit tangent $\mathbf{t}$ will be geodesics if $\mathbf{n} \cdot \text{curl } \mathbf{t}$ vanishes identically: they will be lines of curvature if $\mathbf{t} \cdot \text{curl } \mathbf{t}$ is zero, and they will be asymptotic lines if $\mathbf{t} \times \mathbf{n} \cdot \text{curl } \mathbf{t}$ vanishes identically.

127. Small deformation of a surface. The differential invariants of Chapter XII have also been employed by the author in the treatment of the general problem of small deformation of a surface, involving both extension and shear*. A surface S undergoes a small deformation, so that the point whose original position vector is $\mathbf{r}$ suffers a small displacement $\mathbf{s}$, which is a point-function for the surface; and the new position vector $\mathbf{r}'$ of the point is given by

$$\mathbf{r}' = \mathbf{r} + \mathbf{s}.$$

It is shown that the *dilation* θ of the surface, being the increase of area per unit area, is given by

$$\theta = \text{div } \mathbf{s}$$

* "On small Deformation of Surfaces and of thin elastic Shells," *Quarterly Journal of Mathematics*, Vol. 50 (1925), pp. 272—296.

and that the unit normal $\mathbf{n}'$ to the deformed surface S' is expressible as

$$\mathbf{n}' = \mathbf{n} - \mathbf{n} \times \operatorname{curl} \mathbf{s},$$

the change being due to a rotation of the element of surface represented by the vector

$$\operatorname{curl} \mathbf{s} - \tfrac{1}{2} (\mathbf{n} \cdot \operatorname{curl} \mathbf{s}) \, \mathbf{n}.$$

The first curvature J' of the deformed surface is found to have the value

$$J' = J + \mathbf{n} \cdot \nabla^2 \mathbf{s} - 2\overline{\nabla} \cdot \mathbf{s}$$

and the second curvature K' the value

$$K' = (1 - 2\theta) K + K (\nabla^* \cdot \nabla \mathbf{s}) \cdot \mathbf{n},$$

where $\overline{\nabla}$ and ∇^* are two new invariant operators with properties similar to those of ∇. Inextensional deformation, in which the length of any element of arc remains unaltered, is considered as a particular case.

The above relations are proved at the outset; and the paper then goes on to examine in detail the geometry of the strain. The *extension* in any direction at a point, or the increase of length per unit length of arc, is shown to be $\mathbf{t} \cdot \nabla \mathbf{s} \cdot \mathbf{t}$, $\mathbf{t}$ being the unit vector in that direction; and the sum of the extensions in two perpendicular directions on the surface is invariant, and equal to the dilation θ. The values of the "components" of strain, the existence of principal lines of strain, and a geometrical representation of strain are also examined.

The last two divisions of the paper, dealing with the stresses in a thin shell and the equations of equilibrium, do not belong to the domain of Differential Geometry.

128. Oblique curvilinear coordinates in space. Much of the theory of Chapter XI has been extended by the author to triple systems of surfaces which do not cut orthogonally†. Let the three systems of surfaces be

$$u = \text{const.}, \quad v = \text{const.}, \quad w = \text{const.},$$

the position vector $\mathbf{r}$ of a point in space being a function of the oblique curvilinear coordinates u, v, w. A set of fundamental

---

† "On triple Systems of Surfaces and non-orthogonal Curvilinear Coordinates," *Proc. Roy. Soc. Edinburgh*, Vol. 46 (1926).

magnitudes for the system of surfaces may be defined by the equations

$$a = r_1^2, \quad b = r_2^2, \quad c = r_3^2,$$
$$f = r_2 \cdot r_3, \quad g = r_3 \cdot r_1, \quad h = r_1 \cdot r_2,$$

the suffixes 1, 2, 3 having the same meanings as in Chapter XI. In terms of these quantities the unit normals to the parametric surfaces are

$$\frac{r_2 \times r_3}{\sqrt{bc - f^2}}, \quad \frac{r_3 \times r_1}{\sqrt{ca - g^2}}, \quad \frac{r_1 \times r_2}{\sqrt{ab - h^2}}.$$

Expressions are then determined for the three-parametric *gradient* and *Laplacian* of a scalar function in space, and for the three-parametric *divergence* and *curl* of a vector function.

A formula is also found for the *first curvature* of any surface

$$\phi(u, v, w) = \text{const.}$$

and the properties of the coordinate surfaces are examined in some detail. The intersections of the parametric surfaces constitute three congruences of curves, which are studied along the lines explained in the following Art. Lastly, a simple proof of Gauss's Divergence Theorem is given in terms of the oblique curvilinear coordinates.

129. Congruences of curves. The method of Arts. 103–5, in which a congruence of curves is defined as the intersections of two two-parametric families of surfaces, is not very effective. The author has shown* that a curvilinear congruence is most advantageously treated along the same lines as a rectilinear congruence. Any surface cutting all the curves of the congruence is taken as *director surface,* or surface of reference. Any convenient system of curvilinear coordinates u, v on this surface will determine the individual curves of the congruence; and the distance s along a curve from the director surface determines a particular point **r**. Thus **r** is a function of the three parameters u, v, s; or

$$r = r(u, v, s)$$

and the fundamental magnitudes a, b, c, f, g, h introduced in the preceding Art. are again employed.

* "On Congruences of Curves." *Tôhoku Mathematical Journal,* Vol. 28 (1927), pp. 114–125.

By this method the existence and properties of the *foci* and *focal surface* are very easily established, the equation of the focal surface being

$$[\mathbf{r}_1, \mathbf{r}_2, \mathbf{r}_3] = 0.$$

Moreover, corresponding to the developable surfaces of a rectilinear congruence, are here introduced what may be called the *envelope surfaces* of the congruence. The number of these to each curve is equal to the number of foci on the curve.

Hitherto nothing was known of points on a curve corresponding to the *limits* of a ray in a rectilinear congruence. The existence of such points on a curve is here proved by the following method. First it is shown that:

Of all the normals at a given point, to the curve of the congruence through that point, two are also normals to consecutive curves.

It is then an easy step to the theorem:

On each curve of the congruence there are certain points (called "limits") for which the two common normals to this curve and consecutive curves are coincident: and the feet of these normals are stationary at the limit points for variation of the consecutive curve.

This theorem then leads directly to the definition of *principal surfaces* and *principal planes* for a curve.

The *divergence* of the congruence is then defined as the three-parametric divergence of the unit tangent **t** to the curves of the congruence. The surface

$$\operatorname{div} \mathbf{t} = 0$$

may be called the *surface of striction* or *orthocentric surface* of the congruence. It is shown to have important properties, being the locus of the *points of striction* or *orthocentres*, which are the points at which the two common normals to the curve and consecutive curves are at right angles. The orthocentre of a ray of a rectilinear congruence is the "middle point" of the ray.

The properties of *surfaces of the congruence* (Art. 104) are examined in some detail; and an expression is found for the first curvature of the surface

$$v = \phi(u), \quad \text{or} \quad \psi(u, v) = \text{const.}$$

In terms of the fundamental magnitudes the necessary and sufficient condition that the congruence may be *normal* is

$$fg_3 - gf_3 = g_2 - f_1,$$

which for a rectilinear congruence is simply $f_1 = g_2$. The first curvature of the surfaces, which are cut orthogonally by the curves of a normal congruence, is given by

$$J = - \operatorname{div} \mathbf{t},$$

or, if p denotes the value of the product $[\mathbf{r}_1, \mathbf{r}_2, \mathbf{r}_3]$,

$$J = -\frac{p_3}{p}.$$

The common *focal surface* of the congruences of parametric curves, for the triple system of the preceding Art., is given by

$$p = 0.$$

EXAMPLES XVII

1. If, with the notation of Chap. I, the one-parametric operator ∇ for a curve in space is defined by

$$\nabla = \mathbf{t}\frac{d}{ds},$$

prove that $\nabla \cdot \mathbf{r} = 1$, $\nabla \cdot \mathbf{t} = 0$, $\nabla \cdot \mathbf{n} = -\kappa$, $\nabla \cdot \mathbf{b} = 0$,

$$\nabla \times \mathbf{r} = 0, \quad \nabla \times \mathbf{t} = \kappa \mathbf{b}, \quad \nabla \times \mathbf{n} = -\tau \mathbf{n}, \quad \nabla \times \mathbf{b} = -\tau \mathbf{b}.$$

Also calculate the one-parametric divergence and curl of $\phi \mathbf{t}$, $\phi \mathbf{n}$ and $\phi \mathbf{b}$.

2. If, for a given surface, $\mathbf{l}$ and $\mathbf{m}$ are defined by

$$\mathbf{l} = \frac{\mathbf{r}_2 \times \mathbf{n}}{H}, \quad \mathbf{m} = \frac{\mathbf{n} \times \mathbf{r}_1}{H},$$

show that $\mathbf{l}$, $\mathbf{m}$, $\mathbf{n}$ form the *reciprocal* system of vectors to $\mathbf{r}_1$, $\mathbf{r}_2$, $\mathbf{n}$, satisfying the relations

$$\mathbf{l} \cdot \mathbf{r}_1 = 1, \quad \mathbf{m} \cdot \mathbf{r}_2 = 1,$$

and

$$\mathbf{l} \cdot \mathbf{r}_2 = \mathbf{m} \cdot \mathbf{r}_1 = \mathbf{l} \cdot \mathbf{n} = \mathbf{m} \cdot \mathbf{n} = 0.$$

Prove also that

$$H^2\mathbf{l} = G\mathbf{r}_1 - F\mathbf{r}_2, \quad H^2\mathbf{m} = E\mathbf{r}_2 - F\mathbf{r}_1,$$

and similarly that

$$\mathbf{r}_1 = E\mathbf{l} + F\mathbf{m}, \quad \mathbf{r}_2 = F\mathbf{l} + G\mathbf{m},$$

and show that

$$\mathbf{l}^2 = G/H^2, \quad \mathbf{m}^2 = E/H^2, \quad \mathbf{l} \cdot \mathbf{m} = -F/H^2, \quad [\mathbf{l}, \mathbf{m}, \mathbf{n}] = 1/H.$$

3. In terms of the vectors $\mathbf{l}$, $\mathbf{m}$ of Ex. 2, show that

$$\nabla\phi = \mathbf{l}\frac{\partial\phi}{\partial u} + \mathbf{m}\frac{\partial\phi}{\partial v},$$

so that

$$\mathbf{l} = \nabla u, \quad \mathbf{m} = \nabla v.$$

4. Prove that the focal curve of a family of curves on a surface (Art. 126) is the *envelope* of the family, being touched by each member at the foci of that curve.

5. Show that the focal curve of the families of parametric curves is also the focal curve of the family $\phi(u, v) = $ const.

6. Prove that consecutive parametric curves $v = $ const. on a surface can possess a common normal only where

$$\frac{\partial}{\partial u}\left(\frac{H^2}{E}\right) = 0,$$

and deduce the equation div $\mathbf{a} = 0$ for the line of striction of the family.

7. Show that the foci on a generator of a skew surface are two imaginary points equidistant from the central point, or point of striction. Prove that, as the specific curvature of the surface tends to zero, these points tend to coincidence; and deduce the dual nature of the edge of regression of a developable surface, as formed by the coalescence of the focal curve with the line of striction.

8. The parametric curves are orthogonal, and the curves $v = $ const. are geodesics (div $\mathbf{b} = 0$). If a curve C cuts these at a variable angle θ, its unit tangent is $\mathbf{a}\cos\theta + \mathbf{b}\sin\theta$, and its geodesic curvature is div $(\mathbf{b}\cos\theta - \mathbf{a}\sin\theta)$. Show that this latter expression is equal to

$$-\left(\sin\theta\,\text{div}\,\mathbf{a} + \frac{d\theta}{ds}\right),$$

where $d\theta/ds$ is the arc-rate of increase of θ along C. Deduce the theorem of Art. 126 on a family of geodesics.

9. An orthogonal system of curves on a surface is inclined at a variable angle θ to the orthogonal parametric curves. Show that the unit tangents to the curves are

$$(\mathbf{a}\cos\theta + \mathbf{b}\sin\theta), \quad (\mathbf{b}\cos\theta - \mathbf{a}\sin\theta).$$

Deduce the value of the tangential component of the vector curvature of this orthogonal system, and show that it may be expressed in the alternative forms

$$-(\mathbf{a}\,\text{div}\,\mathbf{a} + \mathbf{b}\,\text{div}\,\mathbf{b}) + \mathbf{n} \times \nabla\theta,$$

or

$$-(\mathbf{a}\,\text{div}\,\mathbf{a} + \mathbf{b}\,\text{div}\,\mathbf{b}) - \text{curl}\,(\theta\mathbf{n}).$$

10. A surface S undergoes a small deformation as described in Art. 127, the displacement $\mathbf{s}$ being of the first order, while small quantities of higher order are negligible. Show that the fundamental magnitudes for the deformed surface are

$$E' = E + 2\mathbf{r}_1 \cdot \mathbf{s}_1, \quad G' = G + 2\mathbf{r}_2 \cdot \mathbf{s}_2, \quad F' = F + (\mathbf{r}_1 \cdot \mathbf{s}_2 + \mathbf{r}_2 \cdot \mathbf{s}_1).$$

Deduce that

$$H' = H(1 + \text{div}\,\mathbf{s}),$$

and hence that the dilation θ is equal to div $\mathbf{s}$. Also show that the unit normal can be expressed as

$$\mathbf{n}' = \mathbf{n} - \mathbf{n} \times \text{curl}\,\mathbf{s}.$$

11. If u, v, w are (oblique) curvilinear coordinates in space, prove that the normal to the surface $\phi\,(u,\,v,\,w) =$ const. is parallel to the vector

$$\phi_1\mathbf{r}_2\times\mathbf{r}_3+\phi_2\mathbf{r}_3\times\mathbf{r}_1+\phi_3\mathbf{r}_1\times\mathbf{r}_2,$$

and show that this vector is p times the rate of change of ϕ in the direction of the normal, where $p=[\mathbf{r}_1,\,\mathbf{r}_2,\,\mathbf{r}_3]$.

12. If the position vector $\mathbf{r}$ of a point in space is a function of the three parameters u, v, w, while a, b, c, f, g, h are the magnitudes of Art. 128, and A, B, C, F, G, H are the co-factors of these elements in the determinant

$$D=\begin{vmatrix} a & h & g \\ h & b & f \\ g & f & c \end{vmatrix}=p^2,$$

prove that

$$p\,\mathbf{r}_2\times\mathbf{r}_3=A\,\mathbf{r}_1+H\,\mathbf{r}_2+G\,\mathbf{r}_3,$$

with two similar formulae. Also show that the unit normals to the parametric surfaces are $\mathbf{r}_2\times\mathbf{r}_3/\sqrt{A}$, etc.

13. With the notation of Ex. 12, if $\mathbf{l}$, $\mathbf{m}$, $\mathbf{n}$ are the *reciprocal* system of vectors to $\mathbf{r}_1$, $\mathbf{r}_2$, $\mathbf{r}_3$ defined by

$$\mathbf{l}=\frac{\mathbf{r}_2\times\mathbf{r}_3}{p},\quad \mathbf{m}=\frac{\mathbf{r}_3\times\mathbf{r}_1}{p},\quad \mathbf{n}=\frac{\mathbf{r}_1\times\mathbf{r}_2}{p},$$

show that
$$\mathbf{l}\cdot\mathbf{r}_1=\mathbf{m}\cdot\mathbf{r}_2=\mathbf{n}\cdot\mathbf{r}_3=1,$$
while
$$\mathbf{l}\cdot\mathbf{r}_2=\mathbf{m}\cdot\mathbf{r}_1=\text{etc.}=0.$$
Prove that
$$\mathbf{r}_1=a\mathbf{l}+h\mathbf{m}+g\mathbf{n},$$
$$D\mathbf{l}=A\,\mathbf{r}_1+H\,\mathbf{r}_2+G\,\mathbf{r}_3,$$

and write down the corresponding formulae for $\mathbf{r}_2$, $\mathbf{r}_3$, $\mathbf{m}$ and $\mathbf{n}$. Also show that

$$\mathbf{l}^2=A/D,\quad \mathbf{m}^2=B/D,\quad \mathbf{n}^2=C/D,$$
$$\mathbf{m}\cdot\mathbf{n}=F/D,\quad \mathbf{n}\cdot\mathbf{l}=G/D,\quad \mathbf{l}\cdot\mathbf{m}=H/D.$$

14. If, with the same notation, the three-parametric ∇ is defined by

$$\nabla=\mathbf{l}\frac{\partial}{\partial u}+\mathbf{m}\frac{\partial}{\partial v}+\mathbf{n}\frac{\partial}{\partial w},$$

prove that

$$\nabla\cdot(X\mathbf{r}_1+Y\mathbf{r}_2+Z\mathbf{r}_3)=\frac{1}{p}\left[\frac{\partial}{\partial u}(pX)+\frac{\partial}{\partial v}(pY)+\frac{\partial}{\partial w}(pZ)\right],$$

$$\nabla\times(P\mathbf{l}+Q\mathbf{m}+R\mathbf{n})=\frac{1}{p}\,\Sigma\,(R_2-Q_3)\,\mathbf{r}_1.$$

Deduce from the former that the first curvature of the parametric surface $u=$ const. is given by

$$J=-\frac{1}{p}\left[\frac{\partial}{\partial u}\left(\frac{A}{\sqrt{A}}\right)+\frac{\partial}{\partial v}\left(\frac{H}{\sqrt{A}}\right)+\frac{\partial}{\partial w}\left(\frac{G}{\sqrt{A}}\right)\right].$$

Also prove the identity $\nabla\cdot\nabla\times\mathbf{F}=0$, where $\mathbf{F}$ is any vector point-function.

15. If a, b, c, f, g, h are the magnitudes of Art. 128, prove the relations

$$\mathbf{r}_2 \bullet \mathbf{r}_{23} = \tfrac{1}{2} b_3, \quad \mathbf{r}_1 \bullet \mathbf{r}_{22} = h_2 - \tfrac{1}{2} b_1, \quad \mathbf{r}_1 \bullet \mathbf{r}_{23} = \tfrac{1}{2} (g_2 + h_3 - f_1),$$

and write down all the corresponding formulae.

16. With the notation of Exx. 12—15, show that, for a *triply orthogonal system* of surfaces ($f = g = h = 0$),

$$D = p^2 = abc, \qquad F = G = H = 0,$$

$$A = bc, \qquad B = ca, \qquad C = ab,$$

$$\mathbf{l} = \mathbf{r}_1/a, \quad \mathbf{m} = \mathbf{r}_2/b, \quad \mathbf{n} = \mathbf{r}_3/c,$$

$$\nabla = \frac{\mathbf{r}_1}{a} \frac{\partial}{\partial u} + \frac{\mathbf{r}_2}{b} \frac{\partial}{\partial v} + \frac{\mathbf{r}_3}{c} \frac{\partial}{\partial w}.$$

The first order magnitudes for the surface $u = $ const. are $b, 0, c$; and the second order magnitudes are $-b_1/2\sqrt{a}, 0, -c_1/2\sqrt{a}$. The first curvature of the surface is

$$J = -\frac{1}{\sqrt{abc}} \frac{\partial}{\partial u} \sqrt{bc}.$$

The second derivatives of $\mathbf{r}$ are given by

$$\mathbf{r}_{11} = \tfrac{1}{2} (a_1 \mathbf{l} - a_2 \mathbf{m} - a_3 \mathbf{n}), \quad \mathbf{r}_{23} = \tfrac{1}{2} (b_3 \mathbf{m} + c_2 \mathbf{n}),$$

and similar formulae, and the derivatives of $\mathbf{l}, \mathbf{m}, \mathbf{n}$ by

$$\mathbf{l}_1 = -\frac{1}{2a} (a_1 \mathbf{l} + a_2 \mathbf{m} + a_3 \mathbf{n}), \quad \mathbf{l}_2 = -\frac{1}{2a} (a_2 \mathbf{l} - b_1 \mathbf{m}),$$

and so on. Lamé's relations are equivalent to

$$a_{23} - \frac{a_2 a_3}{2a} = \frac{c_2 a_3}{2c} + \frac{a_2 b_3}{2b},$$

$$\sqrt{ab} \left[\frac{\partial}{\partial u} \left(\frac{b_1}{\sqrt{ab}} \right) + \frac{\partial}{\partial v} \left(\frac{a_2}{\sqrt{ab}} \right) \right] + \frac{a_3 b_3}{2c} = 0,$$

with similar formulae.

17. With the notation of Art. 129, show that consecutive curves of the congruence can meet only where $[\mathbf{r}_1, \mathbf{r}_2, \mathbf{r}_3] = 0$. This is the equation of the *focal surface*.

18. *For any vector point-function in space, the scalar triple product of its derivatives in three non-coplanar directions, divided by the scalar triple product of the unit vectors in those directions, is an invariant.*

19. For a family of parallel surfaces, the three-parametric function curl $\mathbf{n}$ vanishes identically.

20. If a family of curves on a surface cuts a family of geodesics at an angle which is constant along any one curve, the geodesic curvature of any member of the former vanishes at the line of striction of the latter. (Ex. 8.)

130. Family of curves (*continued*). Some further important properties of families of curves on a surface should here be mentioned. Consider first the arc-rate of rotation of the tangent plane to the surface, as the point of contact moves along one of the curves. We have seen that the direction of the axis of rotation is the direction conjugate to that of the curve at the point of contact (Art. 35). The author has shown* that:

If **t** *is the unit tangent for a family of curves on a surface, the tangential component of curl* **t** *gives both the direction of the axis of rotation of the tangent plane, and the magnitude of the arc-rate of turning, as the point of contact moves along a curve of the family.*

In the case of a family of geodesics **n** • curl **t** vanishes identically, and curl **t** is therefore tangential to the surface. Thus:

If **t** *is the unit tangent for a family of geodesics, curl* **t** *gives both the direction conjugate to that of* **t**, *and also the arc-rate of rotation of the tangent plane, as the point of contact moves along one of the geodesics.*

The *moment* of a family of curves may be defined as follows. Consider the tangents to two consecutive curves at two points distant ds along an orthogonal trajectory of the curves. The quotient of their mutual moment by ds^2 is the moment of the family at the point considered. It is a point-function for the surface; and the author has shown that† *the moment of a family of curves with unit tangent* **t** *has the value* **t** • *curl* **t**. This is equal to the torsion of the geodesic tangent, and vanishes wherever a curve of the family is tangent to a line of curvature (Art. 49). The locus of such points may be called the *line of zero moment* of the family. Its equation is **t** • curl **t** = 0. Similarly the *line of normal curvature* of the family is the locus of points at which their geodesic curvature is zero. Its equation is **n** • curl **t** = 0. And the *line of tangential curvature* is the locus of points at which the normal curvature vanishes. It is given by **n** × **t** • curl **t** = 0.

In connection with a family of parallels the author has proved the theorem‡:

A necessary and sufficient condition that a family of curves with unit tangent **t** *be a family of parallels is that div* **t** *vanish identically.*

---

* "On Families of Curves and Surfaces." *Quarterly Journal of Mathematics*, Vol. 50 (1927), pp. 350—361.

† *Loc. cit.*, § 6.

‡ *Loc. cit.*, § 7.

The quantity div **t** may be called the *divergence* of the family. Thus the characteristic property of a family of parallels is that its divergence is everywhere zero. Further, $-$ div **t** is the geodesic curvature of the orthogonal trajectories of the family; and, if this is zero, the orthogonal trajectories are geodesics. Thus:

The orthogonal trajectories of a family of parallel curves constitute a family of geodesics. And conversely, the orthogonal trajectories of a family of geodesics constitute a family of parallels.

Thus to every family of parallels there is a family of geodesics; and *vice versa*. The expression "geodesic parallels" is therefore tautological, as all parallels are of this nature. And, in connection with the properties of geodesics, the following theorem may also be mentioned *:

If a family of curves on a surface cuts a family of geodesics at an angle which is constant for each curve, the line of normal curvature of the former is the line of striction of the latter.

With the notation of Art. 122 we may define the *flux* of a family of curves across any closed curve C drawn on the surface, as the value of the line integral $\int_0$ **t** $\cdot$ **m** ds taken round that curve. Similarly the value of the integral $\int_0$ **t** $\cdot$ $d\mathbf{r}$ may be called the *circulation* of the family round C. Then from the Divergence Theorem it follows immediately that:

The surface integral of the divergence of a family of curves over any region is equal to the flux of the family across the boundary of the region.

Similarly from the Circulation Theorem we deduce that:

The surface integral of the geodesic curvature of a family of curves over any region is equal to the circulation of the family round the boundary of the region.

And since the divergence of a family of parallels, and the geodesic curvature of a family of geodesics vanish identically, it follows that:

For any closed curve drawn on the surface, the flux of a family of parallels and the circulation of a family of geodesics vanish identically.

---

* *Loc. cit.*, § 7.

Again, if **a**, **b** are unit tangents to the orthogonal parametric curves, we have seen that

$$K = - \operatorname{div} (\mathbf{a} \operatorname{div} \mathbf{a} + \mathbf{b} \operatorname{div} \mathbf{b}).$$

Hence the total second curvature of any portion of the surface is given by

$$\iint K \, dS = - \int_0 (\mathbf{a} \operatorname{div} \mathbf{a} + \mathbf{b} \operatorname{div} \mathbf{b}) \cdot \mathbf{m} \, ds$$

provided the parametric curves present no singularities within the region. We may take a family of geodesics as the curves $v = \text{const.}$ Then $\operatorname{div} \mathbf{b} = 0$. And since the geodesics may be chosen arbitrarily, subject to possessing no singularity within the region, we have the theorem:

The integral $\displaystyle\int_0 \mathbf{t} \cdot \mathbf{m} \operatorname{div} \mathbf{t} \, ds$ *round a closed curve has the same value for all families of geodesics, being minus the total second curvature of the region enclosed.*

If, however, the geodesics of the family are concurrent at a pole within the region enclosed by C, we must isolate this pole with (say) a small geodesic circle C', and take the line integral round both curves. Then, letting the circle C' converge to the pole, we find the limiting value of the line integral round it to be 2π, and our theorem becomes

$$\iint K \, dS = 2\pi - \int_0 \mathbf{t} \cdot \mathbf{m} \operatorname{div} \mathbf{t} \, ds.$$

This formula expresses *the total second curvature of a portion of the surface, with reference to the boundary values of the divergence and the direction of a family of concurrent geodesics, with pole in the region considered.*

This theorem is more general than the *Gauss-Bonnet formula*

$$\iint K \, dS = 2\pi - \int_0 \kappa_g \, ds$$

for the line integral of the geodesic curvature κ_g of a closed curve, which may be deduced from the above theorem as a particular case *.

131. Family of surfaces. We have shown in Art. 119 that, if ∇ is the two-parametric operator for a surface, the second cur-

---

* *Loc. cit.*, § 8.

vature of the surface is given by the formula

$$2K = \mathbf{n} \cdot \nabla^2 \mathbf{n} + (\nabla \cdot \mathbf{n})^2.$$

When we are dealing with a family of surfaces, $\mathbf{n}$ is also a point-function in space; and it should be possible to find a similar formula in which ∇ is three-parametric. The author has shown[*] that, in terms of this operator,

$$2K = \mathbf{n} \cdot \nabla^2 \mathbf{n} + (\nabla \cdot \mathbf{n})^2 + (\nabla \times \mathbf{n})^2,$$

a formula expressing the second curvature of the surface of a family, as a space differential invariant of $\mathbf{n}$. This may be transformed and written

$$2K = \operatorname{div}(\mathbf{n} \operatorname{div} \mathbf{n} + \mathbf{n} \times \operatorname{curl} \mathbf{n}),$$

which expresses K as the divergence of a certain vector.

By analogy with the line of striction of a family of curves, we may define the *line of parallelism* of a family of surfaces as the locus of points at which the normal to a surface is normal also to a consecutive surface. In terms of three-parametric differential invariants, the equation of this line may be expressed[†]

$$\mathbf{n} \cdot \nabla \mathbf{n} = 0,$$

or
$$\operatorname{curl} \mathbf{n} = 0,$$

either of which is equivalent to two scalar equations. With the notation of Art. 128, if the given family of surfaces is the family $w = \text{const.}$, the scalar equations are

$$\frac{\partial}{\partial u}\left(\frac{p}{\sqrt{C}}\right) = 0, \quad \frac{\partial}{\partial v}\left(\frac{p}{\sqrt{C}}\right) = 0.$$

These conditions are satisfied identically for a system of parallel surfaces. Thus:

A necessary and sufficient condition that a family of surfaces be parallels is that curl $\mathbf{n}$ vanish identically.

In closing, we may mention certain other differential invariants of point-functions in space, and point-functions on a surface. In connection with the former the author has proved the theorem[‡].

For any vector point-function in space, the scalar triple product of its derivatives in three non-coplanar directions, divided by the scalar triple product of the unit vectors in those directions, is an invariant.

[*] *Loc. cit.,* § 2.   [†] *Loc. cit.,* § 3.   [‡] *Loc. cit.,* § 4.

If the vector is of constant length, this invariant vanishes. For then the three derivatives are perpendicular to the vector, and are therefore coplanar.

There is a similar differential invariant for a point-function on a given surface; and, in this connection, the author has proved the theorem *:

For the vector point-function **s** *on a given surface, the cross-product of the derivatives of* **s** *in the directions of two unit vectors* **a**, **b** *tangential to the surface, divided by the triple product* [**a**, **b**, **n**], *is independent of the directions chosen.*

Taking the two directions as those of the parametric curves, we see that the value of this invariant is $\mathbf{s}_1 \times \mathbf{s}_2/H$. This function is therefore independent of the choice of parameters on the surface. We have already seen that this differential invariant of the position vector **r** of a point on the surface is equal to the unit normal **n**. And it is easily verified that the same differential invariant of the unit normal has the value $K\mathbf{n}$. If this invariant of any point-function **s** is denoted by $\Lambda(\mathbf{s})$, we have the formula

$$K = \mathbf{n} \cdot \Lambda(\mathbf{n})$$

for the *second curvature* of the surface.

Finally, considering the same invariant of $\phi\mathbf{n}$, where ϕ is a scalar point-function, we easily deduce that the value of $(\phi_1\mathbf{n}_2 - \phi_2\mathbf{n}_1)/H$ is independent of the choice of parameters. And, if **s** is a vector point-functon, by considering $\Lambda(\phi\mathbf{s})$ we find that $\mathbf{s} \times (\phi_1\mathbf{s}_2 - \phi_2\mathbf{s}_1)/H$ is a differential invariant of ϕ and **s**.

---

* *Loc. cit.*, § 4. See also the author's paper "On Isometric Systems of Curves and Surfaces". *Amer. Journ. of Math.*, Vol. 49 (1927), pp. 527—534.

NOTE I

DIRECTIONS ON A SURFACE

The explanation of Arts. 23, 24 may be amplified as follows, so as to attach a definite sign to the inclination of one direction to another on a surface. We define the positive direction along the normal as that of the unit vector

$$\mathbf{n} = \mathbf{r}_1 \times \mathbf{r}_2 / H$$

which we have seen to be a definite vector (p. 4). Then the positive sense for a rotation about the normal is chosen as that of a right-handed screw travelling in the direction of $\mathbf{n}$. Consequently the angle ω of rotation from the direction of $\mathbf{r}_1$ to that of $\mathbf{r}_2$ in the positive sense lies between 0 and π, so that $\sin \omega$ is positive. For any other two directions on the surface, parallel to the unit vectors $\mathbf{d}$ and $\mathbf{e}$, the angle ψ of rotation from $\mathbf{d}$ to $\mathbf{e}$ in the positive sense is then given by

$$\sin \psi \, \mathbf{n} = \mathbf{d} \times \mathbf{e}, \quad \cos \psi = \mathbf{d} \cdot \mathbf{e}.$$

In the case of the displacements $d\mathbf{r}$ and $\delta\mathbf{r}$, of lengths ds and δs, corresponding to the parameter variations (du, dv) and $(\delta u, \delta v)$ respectively, the angle ψ of rotation *from the first to the second* is such that

$$\begin{aligned} ds \, \delta s \sin \psi \, \mathbf{n} &= d\mathbf{r} \times \delta\mathbf{r} \\ &= (\mathbf{r}_1 \, du + \mathbf{r}_2 \, dv) \times (\mathbf{r}_1 \, \delta u + \mathbf{r}_2 \, \delta v) \\ &= H (du \, \delta v - \delta u \, dv) \, \mathbf{n}. \end{aligned}$$

Consequently

$$ds \, \delta s \sin \psi = H (du \, \delta v - \delta u \, dv).$$

Similarly

$$ds \, \delta s \cos \psi = E \, du \, \delta u + F (du \, \delta v + \delta u \, dv) + G \, dv \, \delta v.$$

In particular the angle θ from the direction of $\mathbf{r}_1$ to that of (du, dv) is given by

$$\sin \theta = \frac{H}{\sqrt{E}} \frac{dv}{ds}, \quad \cos \theta = \frac{1}{\sqrt{E}} \left(E \frac{du}{ds} + F \frac{dv}{ds} \right).$$

Similarly the angle ϑ from the direction of (du, dv) to that of $\mathbf{r}_2$ (Fig. 11, p. 55) satisfies the relations

$$\sin \vartheta = \frac{H}{\sqrt{G}} \frac{du}{ds}, \quad \cos \vartheta = \frac{1}{\sqrt{G}} \left(F \frac{du}{ds} + G \frac{dv}{ds} \right).$$

NOTE II

ON THE CURVATURES OF A SURFACE

In the preceding pages we have avoided the use of the terms *mean curvature* and *total curvature* for J and K respectively, because we consider them both unsuitable and misleading. If any justification is needed for the course we have taken it will be found in the following considerations, which show that J is the *first curvature* of the surface, being exactly analogous to the first curvature κ of a curve.

It was proved in Arts. 116 and 119 that, for a curved surface,

$$\nabla \cdot \mathbf{n} = - J \quad\dots\dots\dots\dots\dots\dots(1),$$

$$\nabla^2 \mathbf{r} = J\mathbf{n} \quad\dots\dots\dots\dots\dots\dots(2),$$

$$\mathbf{n} \cdot \nabla^2 \mathbf{n} + (\nabla \cdot \mathbf{n})^2 = 2K \quad\dots\dots\dots\dots\dots\dots(3),$$

the symbols having their usual meanings. If now we wish to introduce a one-parametric ∇ for a twisted curve, it must be defined by

$$\nabla = \mathbf{t}\,\frac{d}{ds}.$$

Then, with the notation of Chapter I for a curve,

$$\nabla \cdot \mathbf{n} = \mathbf{t} \cdot \frac{d\mathbf{n}}{ds} = \mathbf{t} \cdot (\tau\mathbf{b} - \kappa\mathbf{t}) = -\kappa \quad\dots\dots\dots\dots(4),$$

which corresponds to (1). Further

$$\nabla^2 \phi = \nabla \cdot (\mathbf{t}\phi') = \mathbf{t} \cdot (\kappa\mathbf{n}\phi' + \mathbf{t}\phi'') = \phi''$$

and therefore

$$\nabla^2 \mathbf{r} = \mathbf{r}'' = \kappa\mathbf{n} \quad\dots\dots\dots\dots\dots(5).$$

corresponding to (2). Similarly

$$\nabla^2 \mathbf{n} = \mathbf{n}'' = \tau'\mathbf{b} - (\kappa^2 + \tau^2)\mathbf{n} - \kappa'\mathbf{t}$$

and therefore

$$\mathbf{n} \cdot \nabla^2 \mathbf{n} = -(\kappa^2 + \tau^2),$$

so that

$$\mathbf{n} \cdot \nabla^2 \mathbf{n} + (\nabla \cdot \mathbf{n})^2 = -\tau^2 \quad\dots\dots\dots\dots(6),$$

which corresponds to (3). These formulae show that J is exactly analogous to κ as a first curvature, and that $2K$ corresponds to $-\tau^2$. Thus, as the torsion of a curve is frequently called its second curvature, so the quantity K is a *second curvature* for the surface.

This terminology is also justified by the order of the invariants involved. For, on comparison of (1) and (3), it is seen that K is a differential invariant of $\mathbf{n}$ of higher order than J, just as τ is of higher order than κ. And these conclusions are also confirmed by the theorem of Art. 126 on the vector curvature of an orthogonal system. For this theorem shows that J is determined by the curvatures of the orthogonal curves; whereas to find K, it is necessary to take the divergence of the tangential component of this curvature. Hence K is of higher order than J.

The quantity J is not a "mean" at all. The half of J, which is the mean of the principal curvatures, does not occur naturally in geometrical analysis. And K is not the "total" curvature of the surface, any more than τ is the total curvature of a curve. The relation which exists between the areas of an element of a surface and its spherical representation (Art. 87) is not sufficient to justify the title; for the theorem expressed by (48) of Art. 123 shows that J has at least an equal right to the same title. Thus the terms *first* and *second* are more appropriate; and the author believes they will meet with general approval.

INDEX

The numbers refer to the pages

Printed in the United States
By Bookmasters